Electronic Mediations

Katherine Hayles, Mark Poster, and Samuel Weber
SERIES EDITORS

Bodies in Technology

Don Ihde

Electronic Mediations, Volume 5

University of Minnesota Press
Minneapolis • London

The University of Minnesota Press acknowledges the work of Edward Dimendberg, editorial assistant, on this project.

Chapter 1 was previously published as "Bodies, Virtual Bodies, and Technology," in *Body and Flesh*, edited by Donn Welton (Oxford: Blackwell Publishers, 1998), 349–57; reprinted with permission of Blackwell Publishers. Chapter 4 appeared as "Perceptual Reasoning—Hermeneutics and Perception," in *Hermeneutics and Science*, edited by Marta Feher, Olga Kiss, and Laszlo Ropolyi (Dordrecht: Kluwer Academic Publishers, 1999), 13–23; reprinted with permission of Kluwer Academic Publishers. Chapter 7 was previously published as "Technology and Prognostic Predicaments," *AI & Society* 131, nos. 1 and 2 (1999): 44–51; reprinted with permission of *AI & Society* and Springer Verlag London Ltd. Chapter 8 previously appeared as "Phil-Tech Meets Eco-Phil," in *Research in Philosophy and Technology*, vol. 18, edited by Marina Paola Banchettie-Robino, Don E. Marietta Jr., and Lester Embree (Greenwich, Conn.: JAI Press, 1999), 27–38; reprinted with permission of Elsevier Science.

Published by the University of Minnesota Press
111 Third Avenue South, Suite 290
Minneapolis, MN 55401-2520
http://www.upress.umn.edu

Library of Congress Cataloging-in-Publication Data

Ihde, Don, 1934–
    Bodies in technology / Don Ihde.
        p.   cm. — (Electronic mediations ; v. 5)
    Includes bibliographical references and index.
    ISBN 0-8166-3845-4 (alk. paper) — ISBN 0-8166-3846-2 (pbk. : alk.
paper)
    1. Technology—Philosophy.   2. Virtual reality.   I. Title.   II. Series
    T14 .I348 2001
    601—dc21   2001003078

Printed in the United States of America on acid-free paper

The University of Minnesota is an equal-opportunity educator and employer.

12 11 10 09 08 07 06 05 04 03 02   10 9 8 7 6 5 4 3 2 1

For Lael, Rylan, and Barrett

Contents

Acknowledgments

Debts are many—

Linda and Mark are already forefronted. I should like to thank those I have repeatedly conversed with and received critiques from and who have provided opportunities for me, especially my colleagues Bob Crease and Evan Selinger, who share in the technoscience enterprise at Stony Brook; Finn Olesen from Aarhus University in Denmark and Helge Malmgren from Göteborg University in Sweden, for repeated invitations to their departments; and Donna Haraway and Bruno Latour, for their ongoing discussions and responses to my inquiries.

I thank the progression of technoscientists who have visited and been "roasted" in our research seminar and survived: Joseph Rouse, Andrew Feenberg, Andrew Pickering, Albert Borgmann, and Peter Galison. I should also credit the progression of visiting scholars to the Technoscience Research Group who, with their original research projects, always supply me with new ideas and directions.

I also need to acknowledge the office staff in philosophy, who help with all the technical processes arising from changed computers and lost disks.

Finally, I note my appreciation to the editors and staff of the University of Minnesota Press for the necessary support and help in producing the final form of this book.

Introduction
Bodies in Technology

Bodies, bodies everywhere. Philosophy, feminist thought, cultural studies, science studies, all seem to have rediscovered bodies. In part this may be because we have had to do some reflection upon being embodied in relation to the various new technologies that we are encountering in the twenty-first century. Our "reach" has extended now to global sites through the Internet, our experiences have been transformed, we are able to enter cyberspace through the primitive virtual reality engines of the present, and we are tempted to think we can transcend our bodies by the disembodiments of simulation. This book is about bodies in technology. I will investigate several senses of body in relation to our experiences of being embodied. We *are* our body in the sense in which phenomenology understands our motile, perceptual, and emotive being-in-the-world. This sense of being a body I call *body one.* But we are also bodies in a social and cultural sense, and we experience that, too. For most of those reared in western traditions, the female breast is an erotic zone, whereas for many from Asian traditions the nape of the neck is equally or more strongly such a zone. These locations are not biological but culturally constructed, although they are located upon us as part of our bodily experience. I call this zone of bodily significance *body two.* Traversing both body one and body two is a third dimension, the dimension of the technological. In the past perhaps the most familiar role within which we experienced and reexperienced being a body was what I have often called an embodiment relation, that is, the relation of experiencing something in the world through an artifact, a technology. Such human-technology relations are often simple—seeing through eyeglasses, nailing with hammers (Heidegger), negotiating doorways while wearing long-feathered hats (Merleau-Ponty). Perhaps we have forgotten that these simple extensions of the sense of our bodies once posed a problem for our self-identification, and that the new questions raised by virtual reality and intelligent machines have been taken up in earlier eras. Yet few of us today would bemoan the way in which the ditch-digging machine has replaced the muscles of the spade-bearing digger of the nineteenth century. Nor are our occasional encounters with changed technologies encounters only with muscles or bodily might; they are encounters with much wider and deeper notions of who we are. They include the full range of our desires and imaginations.

Technofantasies can begin quite young. My wife, Linda, teaches English as a second language to five- to eleven-year-old children in the Three Village School District. One of the exercises she invented came out of an "R is for _____" question, to which kindergarten boys responded, "Robot!" The girls in the class didn't want this to be the "R" example until Linda indicated that there could be girl robots just as well as boy robots. The first year, the drawings of girl robots were in full color (pink and yellow) with pigtails and tutus, and the boy robots were (black and white) transformer-toy-looking robots with bionic arms sprouting weapons and bodies wearing armor. These technofantasies already reflected social gender-role identifications. As the exercise developed, it became more sophisticated. This year's task was to draw robots that could do something the child would like to have a robot do. This time there were robots that could brush teeth, protect little brother from bad dogs, help with baseball practice, wash the dishes, and clean the floors. The technofantasies assigned to the robots tasks that the children either could not do or did not like to do—robots to the rescue!

On a more adult level, a few years ago I was engaged in a series of e-mail and telephone conversations with one of the editors of *Wired Magazine*.[1] He wanted to know what I thought of the people who seriously expressed the desire to be permanently wired into their computers. At first, I was somewhat incredulous about his claim—who could want such a thing? But there are actual people who desire this symbiosis of computer and body. The persons described who wanted this cyborg existence turned out to have either debilitated social skills or disabled-body-related reasons for the desire. In the case of the social cyborgs, the humans were extreme nerds in the sense that they found negotiating the social intricacies of courtship in an Asian matriarchal society beyond their ability; in the case of the body cyborgs, some extreme physical disability was involved that made a computer synthesis seem preferable to the actual, limited body.

In both examples the technofantasy was based upon the intersection of technologies and human desires in both bodily and social dimensions. Those technologies are still fantasy—imagine the technical difficulties in designing a toothbrushing robot for a wiggly kid, or one programmed to negotiate grandmother's and mother's approval for a bride. The real robot technologies possible today at best approximate insect motion or remain in fixed, closed-system assembly lines, and are indefinitely far from toothbrushing or bride-pricing capacities. Yet these fantasies are actually mild ones compared to the bodily social fantasies now being promoted by techno-utopians. The hype claims that eventually virtual reality (VR) will be better than real life (RL). These stronger fantasies revolve around the notions of hyperreality, virtuality, and virtual bodies and are expressed in the sometimes heard undergraduate statement, "Reality isn't enough anymore."

These examples illustrate how we can "read" or "see" ourselves by means of, through, or with our artifacts. We can—in technological culture—fantasize ways in which we get beyond our physical limitations or our social problems by means of technologies created in utopian imaginations. In this mode of technofantasy, our technologies become our idols and overcome our finitude. But here I seem to be taking a direction that I do not want to take and that I have not taken in earlier works. Unlike our forefathers in philosophy of technology, I am not a dystopian (nor am I a utopian), so I must move carefully in my thinking about technofantasies.

A commonplace temptation might be to associate technofantasy with the genre of science fiction. I am not sure where or when science fiction began—it might have a European origin in the work of Roger Bacon in the thirteenth century. Bacon fantasized marvelous, imaginary machines:

> It's possible to build vessels for navigating without oarsmen so that very big river and maritime boats can travel guided by a solitary helmsman much more swiftly than they would if they were full of men. It's also possible to build wagons which move without horses by means of a miraculous force.... It's also possible to construct machines for flight built so that a man in the middle of one can manoeuver it using some kind of device that makes the specially built wings beat the air the way birds do when they fly. And similarly it's also

possible to build a small winch capable of raising and lowering infinitely heavy weights.[2]

What is remarkable about these thirteenth-century technofantasies is not that then technologically impossible achievements were imagined, but rather that these fantasies should take technological form at all. Flying, lifting of infinite weights, motion at a distance, had all been fantasized previously, but usually through some spirit or demon or witch-like agency rather than a material agency such as technology. One must remember, however, that in the thirteenth century clocks were beginning to pervade European monasteries and cities, trade along ocean- and riverfronts was moved by large customs house cranes, cathedrals were being built by machines utilizing simple physics (although still powered by horses and donkeys), and even the heavens were likened to a clockwork with the first extension of the mechanical metaphors that would dominate early modern science some four centuries later. Technological fantasies about extended possibilities actually well preceded what we much later called early modern science. Leonardo da Vinci, the Renaissance man par excellence, based his visualizations of imagined machines precisely upon Bacon's previously literarily described machines.[3]

Technofantasies are not modern at all, but they take a particular shape in late modernity or postmodernity. That shape is often that of a projected *virtual reality* (an oxymoron) or that which takes place in cyberspace, terms that relate to fantasies about our postmechanical, electronic, and computer technologies of the present. Contemporary technohype sometimes wants to extrapolate, Bacon- or Vinci-like, electronic machines that can take us into these hyper- or virtual realities, well beyond the mundane reality that dominates daily life. I am addressing this contemporary phenomenon in this book.

First, however, I should set some context and give some perspective. My own philosophical career began in what today I would call a generic Continental fashion, that is, as a scholar of European philosophy (or philosophers—I wrote articles about the "classical" European phenomenologists, Heidegger, Merleau-Ponty, and more important and with my first book, Paul Ricoeur). Writing about phenomenology and its related hermeneutic dimensions is not the same as doing philosophy in this style of criticism and analysis. But from my other early analytic side, I remained dissatisfied with exposition, commentary, and even critique. I soon turned to an interest in perception that clearly entails bodies. *Listening and Voice: A Phenomenology of Sound* (Athens: Ohio University Press, 1976) and *Experimental Phenomenology* (New York: G. P. Putnam, 1977), phenomenologies of auditory and visual experience, were my first tries at doing such analyses. What I did not explicitly realize then was that my interest in bodily perception was already linked to a parallel interest in technologies, more exactly instrumentation. My first excursus into philosophy of technology was close behind, *Technics and Praxis: A Philosophy of Technology* (Dordrecht: Reidel Publishing, 1979). What this early history shows is that my taste for materiality and concreteness was already there. I have sometimes called myself a phenomenological materialist. This persistent tendency will be noted here as well.

Perception, bodies, technologies continued to be important interests all through the seventies and eighties—paralleled by interest in technology. But by the nineties the role of science began to capture more of my interest. *Instrumental Realism* (Bloomington: Indiana University Press, 1991) tried to reframe the understanding of science in terms of its late modern technological embodiment in instrumentation, its material incarnation. My take upon science was to try to recall that in addition to mathematizing, modeling, and formalizing a world, science also *perceives* its worlds, albeit through instruments, and that is where very contemporary science meets the above-stated themes of bodies and technofantasies. In the chapters included here on science instrumentation, I show that these technologies proceed from early and simple mechanical and optical devices toward contemporary computer-assisted modeling and, in effect, virtual reality devices. It might seem that popular culture technofantasies concerning hyper- and virtual realities, and the simulation and tomographically constructed science imaging, are convergent trajectories. There is a small echo of precisely this still implicit convergence within the battles of the "science wars."

In a series of books counterattacking postmodernists, feminists, and relativists, the science warriors express worries about science being taken as socially constructed or developing only relative truth or the destruction of objectivity.[4] These worries, however, are often expressed without the slightest self-reflection about how late modern (or postmodern?) science produces or constructs its own imagery, particularly in the state-of-the-art compound instrumentation made possible by contemporary technologies. Has science become virtual without itself knowing it? These, too, are questions that I address in this book.

I became aware of this convergence through a set of serendipitous coincidences. My interest in science instrumentation has been a long one, originally springing from the phenomenologically based insight that bodily perceptions can be embodied through instruments. This partial synthesis between body and instrument makes possible, within phenomenological history, a way to overcome the classical phenomenologists' apparent strong distinctions between a *lifeworld* and separate "worlds of science." A technologically embodied science never leaves the lifeworld. Beginning with what I then termed, and still call, embodiment relations, one can account for a graded set of perceptual magnifications and transformations by following a simple, optical trajectory. Eyeglasses "correct" vision, magnifying glasses (later microscopes and telescopes) enhance vision, and, once this trajectory is followed, an ever stronger set of possibilities for instrument-enhanced vision is opened. Magnification, however, is but a very small part of this story—spatiality, apparent distance, magnification of not only visual objects but reflexively bodily motion occurs, and on and on if one does a truly detailed phenomenology of this phenomenon. One will see echoes of some of this work in the following chapters.

Then a few years ago, my longtime colleague E. Ann Kaplan, director of the Humanities Institute and a film theorist, organized what became known as the Stony Brook Imaging Group. Put simply, we gathered faculty who were interested in related imaging processes— film, television, radiology, computer modeling—to meet occasionally and share working papers on our research. This first round made me recognize that each disciplinary approach, while sophisticated in its use of instrumental techniques and a hermeneutics of reading focal phenomena within the discipline, also was accompanied by areas of virtually blind naïveté about other imaging possibilities. Humanists could not "read" an MRI scan with a brain tumor, even if it were right before their eyes. But neither could the scientists believe, as the Bill Moyers series on truth in television broadcasting showed, that even news broadcasts could be so well spun through the manipulations of the press by President Reagan's spin doctors.

The next year I organized an interdisciplinary graduate seminar, "Imaging Technologies," the first of what has become a progression of these. I spent the first few sessions going into what philosophy of science had to offer for this process (not much), and then brought in skilled image makers and users over the next several weeks. I began with science (an astronomer and a radiologist), then media (a film theorist and a television theorist), and finally several computer modelers. They each had two hours to demonstrate their results, to describe what they thought they were doing, and to answer our questions about the productions they had displayed. Then in the final hour, after our guests were gone, we critiqued and analyzed what we thought had transpired. Not unexpectedly, the scientists turned out to be what I call instrumental realists in the sense that however they manipulated or tinkered with their instrumentation, they were quite sure they were getting what was out there, external reality. By contrast, one might want to say that the film and television theorists were extreme social constructionists in the sense that they thought they were creating a "reality" that they knew to be fictitious or an invention that did not necessarily represent anything external. The computer modelers were hybrids—they knew that the heuristic programs they built into the models were inventions and could be tinkered with, even improved upon; they also hoped that in this process they could get close to some kind of reality. One of the truly ironic twists for the modelers lay in the notion that they could build in degrees of reality, which usually meant higher degrees of ambiguity—the more ambiguous, the more "real."

Much more of interest came out of this seminar, but there was also an overarching unifier: *all image user communities used basically the same techniques to produce their results* in spite of the radically different interpretation of what these results were. Images are produced, constructed. They are made. And to get a good image the maker had to tinker with, tune, the instrument. With contemporary processes, this entails much more than simple focus or light manipulation. It often deals with complicated ways to get contrast and enhancement (one can have straight MRI, MRI with enhancement, or fMRI these days).

It may go as far as producing false color or degrees of exaggeration. The seminar participants saw how each community practiced this, but the users still maintained that they had gotten external reality, or fictional fantasy, or a degree of model realism. We are once again detecting a convergence phenomenon, but from a different perspective.

With this background noted, it is time now to introduce the moves that I make in this collection. I begin with bodies. Chapter 1, "Bodies, Virtual Bodies, and Technology," thematizes some of the observations just made. It is a phenomenology of embodiment and disembodiment, analyzing the ways in which we experience being a body—and of being a virtual body. I try to show that these two experiences are different and then how this difference may be manifest in various ways as we live within our technologically textured lifeworld. Embodiment is a complex phenomenon that envelops both what I term body one, the located, perceiving active body I am, and body two, which is body one permeated with the cultural significances that are also experienced. I debated for quite some time about including what stands here as chapter 2, "The Tall and the Short of It: Male Sports Bodies," in this collection. This chapter is simultaneously highly personal and yet also, as I know from discussions with those who have previously read it, highly generic. It describes some features of what it is like to live out a certain kind of embodiment. Together, these phenomenologically informed chapters set the tone for the rest.

Part II is a conversation within areas of the philosophy of science. Chapters 3 and 4 turn to how science produces its evidence. "Visualism in Science" argues that the practices of producing evidence have, particularly in the West, been cast in a largely visualist mode. I suggest that this is a culturally sedimented practice that in itself is not necessary—and I have received some objection to this claim. But if the "choice" of visualizing is an option, it has been an extremely productive one in that science has found ways to produce an intensely sophisticated visual hermeneutics of things. "Perceptual Reasoning: Hermeneutics and Perception" follows this same strand of thought but introduces the trajectory toward virtuality that I hold is taking place in the contemporary practices of science.

In a slightly broader context, examinations of instrumentation are examinations of material technologies, of the interactions and interrelations of humans with material being. And if, when I began in the seventies, philosophers of science had little to say about this, by the eighties they did. My *Instrumental Realism* tries to make the case that the thinkers included do just that—they look at science practice by taking account of how instruments are employed and how these mutually affect outcomes. While I became acquainted with the work of people like Bruno Latour, Peter Galison, Ian Hacking, and Robert Ackermann because they stood in some degree within philosophical parameters, I did not then realize that a similar strand of materially sensitive sociologists had set out on parallel tracks (in the mid-eighties). Shapin and Schaffer's *Leviathan and the Air Pump* was, of course, the landmark work, wherein the air pump became a crucial "actant" in the formation of the "experimental life."[5] Once I started to read these sympaths, I realized that materiality was breaking out all over. Andrew Pickering deals with "material agency," Bruno Latour with the "non-humans" (door openers and speed bumps), and most radically of all, Donna Haraway with "cyborgs," literal human-animal-technology hybrids.

So by the nineties I found myself engaged in and part of a new conversation, as Richard Rorty terms these things. Part III is a conversation with principals in what today is usually called science studies. Chapters 5 and 6 are friendly debates with Donna Haraway, Andrew Pickering, and Bruno Latour. As I entered this new "discussion group," I realized that the other participants practiced a style of analysis that differed from mine—they all expressed allegiance to some form of symmetry, which, in turn, related to different styles of semiotic analyses. In "You Can't Have It Both Ways: Situated or Symmetrical," originally given at a Danish conference with Donna Haraway and Andrew Pickering, I reflect on these differences and find, once again, that bodies make the difference. Bodies are necessarily situated, which is not to say that one cannot take account of interactions between (my) body and materiality. I come down in these issues with a weighted preference for bodily situatedness. I push this a little further with "Failure of the Nonhumans," which focuses a bit more on the powerful role played by Bruno Latour in all these conversations.

Philosophy of technology reemerges with Part IV. Chapters 7 and 8 are based upon reflective worries on my part, worries about what used to be called rational control. I guess I have to admit that I believe not only that one cannot rationally control technological development, but that framing the question in that way misconstrues the very phenomenon of technologies. Rather, one can enter into the situations, and I argue that the entry ought to be at the research and development stages as well as with the later applied ethics stages, and make nudges and inclinations.

Finally, in the epilogue on technoscience, a term that today has become widely used and could be taken to subsume both science and technology studies, I write about "Technoscience and 'Constructed Perceptions,'" which renews a reflection on the convergence I have noted. This convergence of popular and scientific virtualities is looked at from the perspective of historical and epistemological developments.

Bodies, from Real to Virtual

Part I

Chapter 1
Bodies, Virtual Bodies, and Technology

Late twentieth-century technospeak includes talk about VR and RL within which there is much speculation about virtual bodies. In what should be recognized as a familiar speculative projection, the question is posed as to whether VR will supplant or replace RL. Such techno-worries are not new; in the fifties the question of artificial intelligence (AI) replacing human intelligence was a popular theme; much earlier Luddite-era worries were about machines replacing humans in the productive process—here, if the AI analog were to be retroprojected, one might rephrase the issue as machine muscle replacing human muscle. I even remember a worry during the early days of Masters and Johnson's sexuality investigations when a colleague of mine wondered whether the male member might not be replaced by sophisticated vibrators.

I cite these techno-worries not only because they reoccur with each new advance, but because the success or failure of the pattern projected often remains ambiguous. For the most part, male members have not been replaced, although the increasingly sophisticated techniques of artificial insemination, such as stem sperm cell interspecies implantation, widen the gap between sex-procreation possibilities and nonreproductive sexual activity. Although machine-muscle replacement of much labor has occurred in many areas of production, it has been limited to processes designed within closed systems (such as robotics), whereas AI seems reduced to indirect applications within closed-game situations (such as chess playing), rather than open-context, lifeworld situations. Neither muscle nor mind has reached out into the open world except in human-technology symbiotic forms. In the cases of human-technology symbiotics, both mind and muscle have transformed our worlds. But the worry over VR replacing RL fits into this history of techno-worries, and it begins with the standard form of replacement worry.

I wish to address the themes of virtual bodies in relation to lived bodies (in the phenomenological and Merleau-Pontean senses) and

3

the roles that are played by the technologies that relate to virtual and lived bodies. By way of setting a context, however, I want to begin with several phenomenological exercises in nontechnological examples.

Nontechnological Virtuality

In the process of teaching phenomenology, I have always employed thought experiments utilizing imaginative variations, a Husserlian tactic. One such device, which I used for many years, was to ask the class (upper division undergraduates) to imagine doing something that they had not in fact done, but would like to do, and then begin a critical phenomenological description of this imagined action.

What emerged as a pattern over many classes and years was that the action frequently, even dominantly, chosen was some variant upon flying, with examples often taken from a parachute jump. When asked to undertake the description, a second set of patterns emerged; the classes usually divided between what I call (after R. D. Laing) an embodied and a disembodied mode of the parachute jump. The embodied parachutist described takeoff, attaining altitude, the leap from the open door to experience the rush of wind in the face, the sense of vertigo felt in the stomach, and the sight of the earth rushing toward the jumper. The disembodied describer sees an airplane take off, climb, and sees someone (identified with himself or herself) jump from the door and speed toward the earth. Obviously, in these two cases, where one's body is located in the self-identification is a major issue.

A second step in the analysis calls for varying between the two perspectives. A phenomenological analysis shows a variation between what could be called full or multidimensional experience and a visual objectification of presumed body experience. Where does one feel the wind? Or the vertigo in the stomach? Can it be felt "out there" in the disembodied perspective? The answers quickly show a partial primacy to the embodied perspective. One does not feel the wind in the face or the stomach phenomenon out there; upon being pushed, it is interesting to note that disembodied observers admit (usually) that they don't see their own faces in the quasi-other who is identified as themselves. The full, multidimensional experience gestalts in the here-body of the embodied perspective, whereas the visual objectification out there is spectacle-like.

While both perspectives are possible—and I shall show how deeply embedded both perspectives are in our cultural actions—a dialectic can be shown that gives a quasi-primacy to the here-body with respect to full sensory embodiment experience, compared to the quasi-otherness of the disembodied perspective that nevertheless is a possible perspective that has its own advantages. But the dialectic is weighted with sensory richness given to and within the here-body perspective, which I shall associate with the RL body. The partially disembodied or body as quasi-other perspective is already a kind of virtual body in a nontechnological projection. This form of virtuality is an image-body.

Let us switch for a moment to another popular form of nontechnologized virtual body experience, the much publicized forms of out-of-body experiences. Here the describer, perhaps recalling a situation in an emergency room, claims to have left his or her body, looks down upon this previous body from some floating perspective, and describes the activities going on. The floating perspective is temporary (it can't be known if anyone permanently leaves the body!), and the out-of-body experience returns to a later waking-up experience in one's own body. Phenomenologically, however, this form of experience is parallel to the previously described embodied/disembodied parachute variation, although the popular literature does not recognize this.

In the out-of-body case, the now visually objectified body—the one down there on the operating table as image-body—is identified with "my body," but under the perspective of not being the "now-me" that is implicitly identified with the floating perspective. However, the floating perspective is the now-me and the here-body that is embodied in the new position. I suspect that an interrogator could again note that the multisensory sense of the here-body would locate this in the floating location. The very me/not-me of the body on the table is an indicator of the virtuality of the me-as-other-body in this experience.

The Incredible Lightness of Being (Seated/Lying)

Let us push the here-body, fully sensory body experience a little more. It is precisely the imaginative situation (the students are seated, not unlike the philosopher's standard position) and the already only quasi-conscious (lying down) positions that make the move to "floating" positions more likely. These are, in other words, already only quasi-active situations. Although the subjects in our two variations are not as pathologically impaired as Merleau-Ponty's Schneider, who cannot "be" his active body, they are far from what some of us call the sports body or active body implied as the secret norm of the lived body implicit in *The Phenomenology of Perception*.[1] It is the here-body in action that provides the centered norm of myself-as-body. This is the RL body in contrast to the more inactive or marginal VR bodies that make the shift to the quasi-disembodied perspectives possible.

If I am right about the secret norm of a here-body in action, it should also be noted that such a body experience is one that is not simply coextensive with a body outline or one's skin. The intentionality of bodily action goes beyond one's bodily limits—but only within a regional, limited range. A good example may be taken from martial arts experience wherein one can "feel" the aimed blows even from behind and aims one's own activity beyond any simple now-point. One's "skin" is at best polymorphically ambiguous, and, even without material extension, the sense of the here-body exceeds its physical bounds.

A second ambiguity should also be preliminarily noted, as phenomenological literature has long shown: one can simultaneously experience one's here-body from its inner core while having a partial, but only partial, "external" perception. I can see my hands, feet, part of my frontal visible body from the focal point of my vision.

Combining these multistable ambiguities, one can begin to appreciate how complex the issues of virtuality may become. The opening to a sliding perspective from the multidimensional experience of my here-body toward the image-body perspectives lie within these ambiguities.

Extending the Here-Body

Heretofore, my examples have differentiated a here-body from a virtual or image-body as a disembodied over-there body. The bistability of these two perspectives may be expanded and made complex in many social and cultural activities. For example, in previous work I have shown how a reading perspective that makes a god's-eye view possible gets worked out within European culture in activities seemingly distant from bodies in navigational practice. Europeans locate where they are from a disembodied, overhead perspective. Contrarily, the here-body is made central in South Pacific navigational practice, which makes all motion and direction relativistically referential to the navigator's here-body ("Tahiti is approaching us as the ocean passes our bow"). But in all these cases I have not included nor attended to any technologies as such. Technologies can radically transform the situation, including one's sense of one's body.

This transformation has been descriptively analyzed within classical phenomenology. Heidegger's tool analysis noted that objects, such as hammers, are taken into the ways in which humans project themselves into work practices.[2] When in use, the hammer withdraws as a separate object and is taken into the action being performed. In terms of the language of embodiment, Merleau-Ponty took account of the way in which technologies may be embodied, such as the blind man's cane or the woman's feathered hat. In the first instance, the cane/roadway touch is what the walker experiences—his body is extended through the cane, which becomes part of his here-body experience. In the more radical sense of the hat feather, the sense of her here-body—even without a touch—is extended beyond the outline of the wearer's body. In all three of these examples, one's sense of embodiment changes, although in a reduced and focused way—it is a quasi-extension entailing the here-body. The very materiality of the technology allows this extendability. The tactility that may be had through hammers, canes, and feathers is real but also less than "naked" in its perceptibility. The hardness—but not the coldness—of the nail is experienced through the hammer; the multidimensional "click" of the sidewalk cement and its textured resistance is felt through the cane—but not its grayness; the very draft of wind in a near miss may be felt through the feather—but the extent of the doorway opening remains opaque in its extent. Each of the missing elements can be filled in only by the full bodily sensory awareness that is part of the ordinary experience of the artifact-user's world.

Degrees of Virtuality

In the attempt to overcome these reductions, the newer forms of virtuality take shape. The ultimate goal of virtual embodiment is to become the perfect simulacrum of full, multisensory bodily action. Once this is discerned, one can easily see how far the technologization of virtual reality has to go.

A brief look at imaging media is instructive here. Early technologies, such as the telephone and the phonograph, or still photography and silent motion pictures, were monosensory as either audio or visual media. Phenomenologically, however, users constantly experience media multidimensionally; thus the monosensory quality of these media easily reveal the technological reductions that are simultaneous with the more dramatic amplifications and magnifications that occur within the audio or visual media. One could not see the telephone caller, nor could one hear the speaker in the silent movie. An early response to this reductive limitation was to make the monosensory dimension richer—silent movies called for more mime, more gestural significations, producing a kind of visual exaggeration, and the singer on the phonograph record could exaggerate vocal gesture.

However, a richer technological trajectory for media soon followed. Within a few decades movies (1889) became talkies (1927), and the medium became audiovisual. Now bidimensional audiovisual media are effectively the norm of many communicative media (cinema, television, camcorders, most multimedia presentations including computers) and have been for nearly half a century. The audiovisual has become deeply sedimented in our seeing/hearing and is taken for granted in our experience.

Far less success, although again noted from the times of very early media developments, has been attained with respect to tactility, kinesthetic, gustatory, or olfactory dimensions. Theaters with shaking floors or "smellovision" have been attempted, but far more dominant are projected synesthetic attempts to induce the multidimensional experience; synesthetic vertigo in Cinerama or maxitheaters and auditory overkill with Dolby or sensurround sounds are the equivalents to early mime and gestural compensations that remain within the possibilities of the audiovisual.

I take account of this recent history (which actually begins at the very end of the nineteenth century, accelerating in the twentieth) because it relates intimately to technologized attempts at virtual embodiment. Within this history the ambiguity of the embodied/disembodied or here-body/image-body continues. If, for example, we begin with the here-body variant, it can quickly be seen that media employ the same perceptual referencing that I noted in Pacific navigation. The viewer is seated in the theater in a fixed position, but the imaged world is in motion and referentially aims at the viewer. The tracks of the roller-coaster appear to slide under the viewer's seat until the apogee is reached, then tilt to show a downward, vertiginous acceleration; the synesthetic fall begins. Here the imaged version of "Tahiti comes to me" is the roller-coaster descent that rushes toward me in screened quasi-realism. In contrast, if one were to see a shot of screaming riders on a roller-coaster from "out there," the synesthetic effect either disappears or is muted (occurring, if at all, in the here-body position). The other is seen to have the vertigo. The virtual realism is enhanced when the imaged environment refers back to the seated viewer.

All these effects presuppose the privilege of the mostly motionless (seated) viewer. They are the technological equivalents to the assembly line or other relatively closed system within which the technologies may perform the limited actions that are the machine worlds we know. The viewer must move into the equivalent of the screen-theater environment to become an actor. This becomes a possible trajectory for an (embodied) virtual body.

Earlier attempts to inject more interactivity may be noted in the range of simulators that originated with military technologies. During the early days of World War II, it was noted that if fighter pilots could somehow survive the first five air battles, their chances of long-term survival became much higher. If, therefore, one could simulate what needed to be known before actual battle, the chances for survival might improve. This plan resulted in the first Link Trainers, which allowed dogfight practice to occur inside an early virtual reality situation. The trainer had a projected scene on the cockpit window, the entire trainer moved with the control stick, and sounds and as much realism as possible were injected. These effects are highly sophisticated in contemporary military and industrial simulators, which are used to train pilots, tank personnel, and drivers. Roadways, runways, unexpected critical situations, all rush toward the participant, the (seated) viewer, with as much realism as the imaging process can summon. Here, though, the situation is much more interactive because simulation controls call for actual bodily action, which enhances the synesthetic effect and adds at least restricted tactile and kinesthetic aspects to the experience.

Commercial entertainment applications followed—video games in parlors or homes, with mostly boys and their dads plugged into projected scenarios of aliens or other enemies. My son and I both play Flight Simulator, with some interesting differences. We both prefer the Lear jet, compared to the Cessna or the WWI biplane, due to its high maneuverability and capacities that are built into the program. But our attitudes are different. I was stressed when I got lost over an Iowa cornfield and ended in a crash. Mark enjoys deliberately smashing into Chicago's tall buildings on takeoff and then repeating, with the disembodied perspective, the scenes of the crash from the quasi-distance of that perspective as the plane parts fall to earth. "Lives," after all, are infinitely repeatable in virtual reality. I wonder if Nietzsche thought of this possibility in his theory of repeatable infinite choices? Or is the doctrine of eternal recurrence simply an anticipation of video-game culture?

Yet phenomenologically, this admittedly more actional technological space is but a small step from previous more passive audiovisual situations. The flyer remains seated, and the screen-world back-projects the framed action to the viewer. Action remains minimal in the movement and synesthetic amplification of the body through the joystick. It is all hand-eye coordination, enhanced in the context of hyper-graphics, sound effects, and synesthetic amplification. (We remain far from "virtual food," and as early smellovision showed, the compressed time frames of theaters are not enough to remove and add smells in sequence as in the more durational spans of ordinary life.)

In terms of present virtual simulations, we have but one more step to go—the step to the technologically "wired" body cages that include "face sucker" goggles, gloves, and perhaps even strapped-in motion cages as shown in the film *Lawnmower Man*. These developments display a slightly different tactic with regard to here-body referencing. One of the shortcomings of other simulation forms has to do with the framing, which has always differentiated the artificiality of the technologically imaged from the wider world. The video game, even with a very large screen, remains framed by the screen. While the screen may even surround the viewer in maximal settings, the quasi-depth of the screen remains a detectable film artifact. Everything is in front of the participant. The technological cage follows a direction developed earlier in sensory-deprivation experiments. The participant is surrounded with the technological cocoon, which is equivalent to making one's "world" much closer and more encompassing. The mini-TVs directly in front of the eyes, the bodysuit, the wired gloves, all enclose the participant in the up close environment of the technologically encased envelope from the RL world. This enclosure, however, is neither neutral nor transparent—its vestigial presence may produce a sense of both unreality and disorientation, of a kind of claustrophobia known to produce nausea in some participants. Again, the effects are similar to those of sensory-deprivation experiments in the seventies, but whereas the deprivation experiments damped out the body/world differences, the VR version makes the world a hyperworld.

Once again, phenomenologically, the VR cage remains simply a different degree of virtuality of the open but framed version in the video game. It does introduce—so far primitively—tactility and kinesthetic effect into the medium, and thus is a step beyond the merely audiovisual, seated context. But even with this greater degree of actional possibility, most VR programs—except for the most expensive and sophisticated—lack the feedback found in full bodily engagement. Actions taken usually lack the sense of contact, and the price remains that of closing, to a greater degree, the openness of the RL world in order to attain its effect. VR vertigo remains an insulating vertigo. It remains VR theater. It is, however, a very special kind of theater. Its audience is individual (although there are settings in which multiple players engage); its world is programmed with the usual logic trees of choices (none of which provides the ambiguities or openness of RL) that, while complex, do not adapt to learning novel possibilities.

Here we reach one horizon from which the original techno-worry fed. Can VR replace RL? Only if theater can replace actual life. Only the bumpkin rushes to the stage to rescue the maiden from the villain, but the late twentieth century is apparently filled with willing bumpkins! Like theater, VR developments contain devices that enhance the VR experience and distract from RL contexts.

Imaging the Technologically Polymorphic Possibilities

"Morphing" things in the VR world and the use of hypergraphic techniques that use bright colors and lighting effects are part of the theatricality of current VR. It is interesting to note in passing that the contemporary version of the long-standing automaton worry of

Cartesian philosophy has been changed. The older Cartesian worry was whether or not we could be deceived by a cleverly contrived robot, a look-alike. The new worry is whether hyperreality is such that "reality isn't enough anymore." In the film *Lawnmower Man*, the eponymous character has a VR sexual encounter with his girlfriend in which both he and she are morphed into fantastic shapes and interactions—projected in this case as image-bodies in the second perspective for the movie viewers—ending with an imaged Georgia O'Keefe–like "orgasm" climax. In the end, however, how different is this from other movie techniques that add romantic music, off-center shots, and suggestive body parts? The difference is simply one of nonhuman morphed shapes and the suggestion that the technologies make hyper-sex different. VR is a latecomer theatrical development that forefronts techno-imaging.

Imaging, now technologically embodied, makes polymorphy—particularly of visual shaping—a forefront phenomenon. I have already noted the morphing of human body shapes in *Lawnmower Man*, but there is a continuum of variations, all of which do different degrees of morphing with respect to the VR/RL portrayal of bodies. *Roger Rabbit* refined older cartoon/human interaction by making the fictive being three-dimensional and thus presumably more lifelike. Here the cartoon/human (image) world is a hybrid. Farther to the "right," and into a kind of presumed realism, are the newer computer effects that make presumed real entities hyperreal. *Jurassic Park* computer-generates some very lifelike dinosaurs, and more recently, *Twister* magnifies tornadoes into hyperwinds. To the left are the already unreal morphings that either show realistic-looking oddities, such as the parasitic alien animals in *Aliens*, or abstract, vaporous (spiritual?) forms, such as the high-speed travel morphing in science fiction—*Stargate*, warp speed in *Star Trek*, etc.

In one sense, morphing is more a revival than an innovation. It is a return to premodernity in the sets of cultural beliefs that things could actually transmute or metamorphose. Devils inhabiting human bodies, human witches taking on animal shapes, the possibilities of monsters, prodigies, and freaks—all were premodern morphs. Imaging, particularly technically sophisticated visual imaging, reinvents this polymorphism of bodily possibilities. Its culture is a bricolage, where boundaries and distinctions are blurred, parts interchanged, hybrids produced.

In this context, the body, bodies, are but one target. Morphing, rapid exchange of the embodied and disembodied perspectives, on to the horizons of gender blurring (these are enhanced in the more reductive and still mostly monosensory Internet contexts where linguistic morphing is the norm), all are part of the same cultural movement. In one respect all this could be harmless, simply a new variant of ancient fascination with the bizarre and with curiosities. It echoes the earliest history of precinema camera obscura theater wherein paying customers would come to see the magic lantern back-project images of devils and ghosts upon the screen, or perhaps, it even goes so far back as to reach to Plato's precinema cave wherein images of images were the only "realities" for the dwellers prior to Platonic sun therapy.

In each case, however, the illusions are harmless only so long as the experiencer knows the difference between theater and daily life, so long as the one living RL does not become bumpkinlike and take VR as the real. This would be the Platonic solution. For Plato it was the liberation from the cave and the emergence into sunlight that taught the difference. But in a broader, more phenomenological sense, both RL and VR are part of the lifeworld, and VR is thus both "real" as a positive presence and a part of RL.

Virtual Bodies as Technofantasy

I want to conclude this foray into virtual bodies with something of an epistemological moral: VR is a phenomenon that fits neatly into our existential involvement with technologies. Here the question is a deeper one involving our desires and fantasies, which get projected into our technologies.

Concerning the existentiality of our technologies, particularly those that implicate embodiment, I have made this point:

> The direction of desire opened by embodied technologies also has its positive and negative thrusts. Instrumentation in the knowledge activities, notably science, is the gradual extension of perception into new realms. The desire is to see, but seeing is seeing through instrumentation. Negatively, the desire for pure transparency is the wish to escape the limitations of the material technology. It is a pla-

tonism, returned in a new form, the desire to escape the newly extended body of technological engagement. In the wish there remains the contradiction: the user both wants and does not want the
technology. The user wants what the technology gives but does not
want the limits, the transformations that a technologically extended
body implies. There is a fundamental ambivalence toward the very
human creation of our own earthly tools.[3]

This contrary desire applies with particular pathos and poignancy to
desires and fantasies of body. The VR/RL distinction gets blurred,
seemingly crossed, or fantasized about, but there is another aspect of
technology/bodies that is more than play, more than theater and ordinary life. That is, of course, the way in which we increasingly literally incorporate (pun is deliberate) technologies.

Prostheses, from the simple tooth crown to an artificial limb, are
base-level examples. Technofantasies romanticize prosthetic amplifications as bionic that theatrically are precisely the actualization of
the existential contradictions concerning technologies. *Robocop, Bionic
Man,* and *Terminator* all have more-powerful-than-human prostheses,
which in fantasy nevertheless function as freely and spontaneously
as one's lived body. But actual users of prosthetic devices know better—prostheses are better than going without (the tooth, the limb,
the hand), but none have the degree of transparent, total "withdrawal"
of a tool totally embodied. All remain simply more permanently attached ready-to-hand tools.

Yet when one's body fails or is irreparably injured, or parts of it
are removed, the prosthesis becomes a viable and helpful compromise.
It's just that we apparently can't have both the technological empow-
erment and the perfect transparency at the same time.

But the more extreme the situation, the stronger the fantasy may
become. Some years ago I had a number of discussions with a person
I have never met face-to-face about the fantasized desire on the part
of some people who wish to be actually and permanently wired to
their computers. In discussions by telephone with some of these individuals, I learned that one conversant was a severe rheumatoid
arthritic—and I understood precisely, because my own mother had
the same disease. Her body had become so sclerotized that she was
confined to a wheelchair; to eat, someone had to place a spoon into
the fixed fingers of her hand; talking, of course, was possible, but had

she been able to use a computer she would have had to hunt and peck with a pencil, a letter at a time. The fantasy of bypassing an arthritic body, of becoming embodied anew, even through a computer, was understandable. But it is understandable precisely on the condition that the free-flowing, active "sports body," which remains the secret norm of Merleau-Pontean bodily intentionality, is no longer the living possibility of being open to the world.

VR bodies are thin and never attain the thickness of flesh. The fantasy that says we can simultaneously have the powers and capacities of the technologizing medium without its ambiguous limitations, so thoroughly incorporated into ourselves that it becomes living body, is a fantasy of desire. And when we emerge from the shadows, effects, and hyperrealities of the theater into the sunlight in the street, it is not Plato's heaven we find, but the mundane world in which we can walk, converse, and even find a place in which to eat.

Chapter 2
The Tall and the Short of It
Male Sports Bodies

As noted in the introduction, I wondered if I should include this chapter in the context of bodies in technology since there is no direct relation to technologies here. Yet this chapter addresses with some specificity the body one/body two distinction that runs through the discussions, and even more it engages the extremely important discussion of bodies in contemporary feminist discourse, so it seemed appropriate to include this somewhat autobiographical and personal discussion. Contemporary feminism, in fact, has probably done more to reopen the issues of embodiment than any other single strand of recent thought, and issues of gender are related to the question of bodies and extend to the issues of bodies in technology.

The challenge for this chapter comes from Susan Bordo, who complains in "Reading the Male Body" that males themselves have written little about the male experience of one's body: "Men have rarely interrogated themselves as men ... for they generally have not appeared to themselves as men, but rather as the generic 'Man,' norm and form of humanity. . . . When men problematize themselves as men, a fundamental and divisive sexual ontology is thus disturbed."[1] I am responding to this challenge. A second inspiration for this article comes from Iris Young, most particularly from her "Throwing Like a Girl," an early essay to which she later responded retrospectively, noting the experience of her teenaged daughter's somewhat changed situation.[2] As a reader of much feminist literature, beginning with feminist critiques of science and technology—my usual focal interest—but also regarding "bodies" that often enter into phenomenological contexts, I find a set of issues that revolve around how one experiences embodiment.

Within a phenomenological history, the issue is one that falls between two contrasting godfathers, Merleau-Ponty on one side, and Michel Foucault on the other.[3] One might characterize the Merleau-Pontean perspective as a form of phenomenological materialism insofar as his concept of the lived body (*corps vécu*) is one that holds that the active, perceptual being of incarnate embodiment is the very opening to the world that allows us to have worlds in any sense. At bottom, the anonymity of the active, perceiving bodily being he seeks to elicit could be said to be both preconceptual and precultural, and without this sense of body, there is no experience at all. And it is certainly developed and described from a first-person perspective. In contrast, Foucault's body is thoroughly a cultural body, often described and analyzed in a third-person perspective. The body objectified by the medical gaze in the clinic, the body of the condemned in the regicide, and the subjection of bodies within all forms of discipline are culturally constructed bodies. Insofar as there is experience, it is experience suffered or wrought upon human bodies.

I have previously used the term "body one" to refer to the bodily experience that Merleau-Ponty elicits, particularly in *The Phenomenology of Perception*,[4] and "body two" for the culturally constructed body that has echoes with a Foucauldian framework. However, what I term body two is the cultural body as experienced body.[5] But the problem, as it often takes shape in feminist literature, is the combination or noncombination of these perspectives within the experience of the writer. Young is particularly good at combining the sense of acting and of being seen in the ambiguous transcendence of throwing like a girl, the self-pleasure and the social construction of breasted being, and the obviously unique self/other experience of pregnant embodiment. These themes all announce the ambiguities that keep any clear line of demarcation between body one and body two from being drawn. Young thus combines, with at least partial success, the notions of lived and culturally experienced body.

Young also indirectly led me to understand, within Merleau-Ponty's *Phenomenology of Perception*, something that I had not previously seen. Merleau-Ponty sets up a dialectic between what could be called a normative body experience and the pathological experience that is only indirectly noted in his famous Schneider. Young, in her critique in "Throwing Like a Girl," shows that Merleau-Ponty's body experience is abstracted from gendered bodies, and thus implicitly may be describing the focally "masculine," since he does not capture the ambiguous transcendence of female embodiment. That point is granted— but *what is the masculine body experience that is inherent in the corps vécu?* I argue that the emphasis upon perception within actional situations, the transparency of the normal body, the sense of "I can" that occurs in *The Phenomenology of Perception, is open to being secretly a normative sports body.* This slide from a normative, active, and externally focused body to the athletic body allows certain body two aspects to come into play. This healthy, implicitly athletic embodiment contrasts with the debilities of Schneider—but also by extension with virtually any other form of unhealthy, or even less than well-conditioned sense of body, as well as any ambiguously transcendent female embodiment. The normative body implied is certainly neither the feminine one that Young describes, nor one of an arthritic, aging body, nor even of a clumsy, unathletic body. It was this understanding that I began to have as I read Merleau-Ponty through Young's critique.

Growing Up Male: A Retrospective

This, however, provides the opening to something quite unremarked upon in much body literature, which relates to male embodiment and which may reopen the issues of body one and body two in ways that find resonance with gender differences and the socialization of how one lives one's embodiment. I shall make my attempt to "reveal male bodies" precisely in parallel with Young's earliest article and her subsequent retrospective via her daughter. In my case, I shall begin with retrospective observations.

I am a recycled father in the sense that I now have a young son with my second wife; he was born in my mid–middle age. Hearing of his experiences in elementary school revived long-ignored memories of growing up male. Initially, I thought that his experiences would be very different from mine, precisely because his circumstances are so very different. I grew up in a small, largely Scando-Germanic farming community in Kansas; I went to a one-room school and a small high school before finally leaving home to attend a university. Mark is growing up in a densely populated suburban community on Long Island; he attends an intellectually gifted (IG) program in a school system in which the values of the university, the Brookhaven National Laboratory, and the high-tech industries of this region dominate. The intensive and largely science-focused values of the system are evidenced in the fact that the local high school produced the largest number of Westinghouse semifinalists from any single school in the nation last year. All of this had led me to believe that Mark would not have to experience anything like what I experienced as a boy child.

But I was wrong! And, as I shall try to show, part of the reason relates quite directly to the way masculine embodiment is (still) socially embedded. This social embeddedness is very long lasting, historically, and is expressed similarly across radically different social contexts. Mark was usually the youngest boy in his class and at the lower end of the class height spectrum before his adolescent growth spurts began. His peer group contained some quite large boys, but when teams formed at recess, the boys from the IG classes were usually well outsized by the boys from the regular classes. The bumper stickers that proclaim "My child is an honors student at _____ School" are countered by "My child beat up your honors student" on other bumper stickers. There had been such "wars" between the IG classes and the regular classes in years previous to Mark's entry to this program, but these have been ameliorated under a new principal. Mark's IG peers' parents are mostly professionals (doctors, lawyers, professors, scientists, administrators—together with a goodly number of students from graduate student families, including a significant number of Asians). It should be noted in passing that much that goes on in the boys' culture regarding being macho and the like is explicitly disliked and discouraged by the parents. In the contemporary student culture terminology, this means that there are basically only nerds and geeks— certainly no "dirtballs"—but there remains a subclass of jocks within the IG group. I had assumed (wrongly) that this special group selectivity would modify the pain of growing up male that I had experienced.

Those who are parents will easily recognize the next dimension: cliques or gangs. Within the class there are "first," "second," and "third" stringers (the terminology comes from actual conversations), and these obtain for both boys and girls, separately at this stage, of course. Being aware of this, Mark sought the prestige of being associated with the first stringers. This aim was difficult to attain for two simple reasons: size and athletic prowess. He skied, swam, and sailed better than most, but these are the wrong sports for most of the school year. Soccer, baseball, and football dominated the recesses. He managed, but sometimes with difficulty, and often by using alternative skills. His musicianship was well recognized, and the IG classes dominated the orchestra and, to a lesser extent, the band. And he had honed the skills of comedy such that he played in a recognized socially positive role.

Now a time warp, back to my experiences in Kansas during my boyhood. I was the "shrimp" even within my family. My father was 6'2", my younger brother is 6'3", and even my older grown son is also 6'3"—I am only 5'11". As a younger boy, not only was I shorter than most of my classmates (in this Scando-Germanic community), but also a skinny lightweight. (Those who know me from much later post-Ph.D. times will probably find this amusing!) When it came time for choosing players for baseball or especially for football or basketball, I always felt the pain of being one of the last boys chosen by the larger and older jock captains. High school was even worse. I was simply too light to be of much use in football, and although I had grown a bit taller, I never made more than the second team in basketball. In the end, I chose an alternate means of recognition—many Midwest-erners will recognize this—music, and I became the drum major of the marching band. That was sufficient to keep me at the edges of the first string, but barely. All of this was also painfully conscious within adolescent experience.

The above suggests enough similarities between boyhood in the late forties and early fifties and the differently contexted nineties of my son's experience. In both our childhoods, there is a rather direct connection between height and athletic prowess, read quite directly off the body. (There are nuanced exceptions to even this connection, in that there can be tall boys lacking athletic prowess, and short ones who have it and thus can to some degree compensate within the jock domain by sheer aggressiveness.) From here I will begin to focus upon embodiment in these two frames. It is obviously the "tall" of it, com-bined with athletic prowess, the sports body, that plays a crucial role. To set the stage, I bring up some phenomenological variations actu-ally experienced in later life. The Scando-Germanic setting of Kansas, life among tall people, gets repeated today when I travel to Germany and Scandinavia—there I am at most average or just slightly shorter than average height. A subtle shock occurred when I moved in my early career to Long Island and Stony Brook—I suddenly found that I was quite tall compared to the average populace. The genetically determined population height here is more dominantly Eastern and

Southern Euro–American, plus a very large Asian-American population. In my first visits to South America, I was a virtual giant compared to the averages there. To look up or to look down in relation to the other is an experienced bodily comportment that carries many subtle significations, including gender relations, given the bimorphism of the human species. And while there is a ratio of more direct to less direct reading-perception of relative height (and prowess) between childhood and adulthood, it remains one variable within interbody perceptual situations.

One can turn to another set of variations that complicate the situation even more. While a graduate student, one of my admired professors was Paul Tillich. I took three courses with him, and in each case the location was a very vast lecture hall at Harvard—usually with three to four hundred persons in attendance. He was something of a cult figure in the intellectual climate of the time. Later, I finally met him face-to-face when asked by my colleague to serve him a glass of sherry at a cocktail party after a lecture at MIT—I gladly did, and found to my amazed shock that I was looking down upon someone whom I had always thought something of a giant—from the perspective of being in an audience looking up to him behind a lectern. Similar experiences have since occurred with several other famous and notable persons. Social size may differ from bodily size within these variations. The aura of size that does not correspond to physical size is clearly a social-psychological phenomenon. The contrary phenomenon—a Napoleon complex—may also be noted. Short persons deliberately asserting power not through size but through styles of assertion (beware short administrators!) are familiar. All of these variables are obviously contingent since, as Sartre has so often pointed out, one chooses the style with which one lives one's circumstances—one can live size aggressively, meekly, etc. I am not arguing body determinism, only persistent social patterns that do include bodies.

I am pointing out that in male socialization, particularly in childhood, the role of body size and prowess is poignantly important. Much is read directly off body size and skill. Is it different from female experience? (I do not pretend here to speak for female experience, except referencing through what I read in the literature.) It is probably the case that boys are rarely concerned deeply with body parts. Here the external focus echoing Merleau-Ponty seems to hold. Boys do not have to worry about breast size (but that occurs for females in adolescence and thus later than the frame I have selected); not even penis size, for the most part, plays any important role. Bordo has pointed out the hiddenness of the penis in dress, display, and other social situations (now modified slightly with the ascendance of beefcake phenomena and the nude or near-nude hunk): "The penis remains private and protected territory. The woman's body has become increasingly common cultural property. . . . the penis has grown more, not less, culturally cloaked over time. Is it possible to imagine a *Newsweek* cover on testicular cancer, illustrated comparably to the cover on breast cancer?"[6] In my own youth, I can remember only one occasion in which any size comparison was actually made. Boys do not seem to develop the same awareness of "the gaze," the feminine response to which includes crossed legs, motions to cover or hide body parts, etc., as if the gaze were a "penetration" into areas of privacy. At least, for boys, "the gaze" is relatively muted. Boys do, in adolescence, learn the gendered gestures of sitting—not with crossed legs, but by lying back with spread legs. (A colleague has pointed out one of Norman Rockwell's pictures that depicts a skinny youth aggressively handling a pair of "Indian clubs" in front of a mirror with a magazine lying open on the floor showing a picture of Charles Atlas in one of the ads in which cartoons point to a movement from the "90-pound weakling" to the muscular athletic body of the post-Atlas course. Rockwell could hardly be termed a "Lacanian"!) Does this imply a "gaze" of some sort within male socialization? Its variation, if not one of protective hiding, is nevertheless a response to a felt cultural norm.

There is another equivalence that, in some ways, is possibly worse. If the implicit sports body that can connect to the Merleau-Pontean tradition holds, it is largely what is exterior and actionally outward directed that counts more: how far I can throw the baseball is much more important than size of biceps as such; how fast I can run the dash is more important than the appearance of my calves; how well I can execute the jackknife dive is more important than how my buttocks look. (At a high school reunion some decades after graduating, my former coach had one memory of me—the second stringer who once, just before the bell, lofted the basketball for a goal from the center court [and we lost anyway]. It was my only moment of glory, according to him. Of course, my memory of him was not so kind either—I remembered him as the person who largely deprived me of a good basis in geometry and algebra since his teaching method was to assign the class problems to do and then gathered the first team around his desk to plot strategy for the next game.) By late secondary school I had learned to dislike the whole jock culture of the time.

But if the external performance remains "short" of the ideal expectation, is not the sense of failure or "falling short" a similar social phenomenon—indeed, one that affects the whole sense of who I (bodily) am? It is not some part that is deficient; it is the whole (bodily) self. Such experiences lead to multiple strategies. Some of Mark's companions withdrew into total geekhood, including one very large but unathletic boy. Some excelled in recognizable, acceptable alternatives, such as music, computer skills (now a big social factor among boys), and at least in the IG program, intellectual brilliance. (The anti-intellectualism of my Kansas home area was comparatively quite powerful, so other than music and drama, I kept alternative pursuits more secret.) Some used their skills in storytelling, making macho comic strips, or playing computer games, and the like. All such multiple compensation strategies helped to create a socially constructed aura that partially—but only partially—compensated for "deficient" embodiment. Size combines with performance in this hermeneutic of shortcoming.

There is a developmental aspect to this as well. One hoped that with time, the height differences will diminish—tale after tale of late growth is common among male adolescents—and sometimes sports acuity also comes later, while abilities other than the direct reading of size and acumen off bodies also accelerate. But before this happens, there is the pain of the hybrid body-social construction that occurs in young male culture. Long since my boyhood days, it still remains the case that the sixth graders dominate the lower graders on the bus, the bullyboys rule the recess times, and the jocks are the first to receive female attentions. (This, no doubt, tells about something more general than single-gender formation.) If one is short and lacks the presumed ideal agility, one suffers through boyhood.

This is the stuff from which boyhood coming-of-age novels, movies, and stories are made. Only some boys, presumably those at the top of the heap (tall, athletic, strong—and handsome), ever have the ease that too often is attributed to the entire class of males. For most, it is struggle, the finding of alternatives, even the dreaming of revenge in some other form that characterizes boyhood. What has appeared in my analysis so far is a movement of convergence. Boys growing up experience insecurities, tensions between social expectations and bodily realities, and variations upon recognizable gender themes. And while the male experiences display differences from the descriptions females give of similar phenomena, the differences are not so stark as to yield unrecognizable human experiences as variations within wider human embodiment spectra.

The Mauss Illusion and Gender Reciprocities

In anthropology a well-known phenomenon is the Mauss illusion. Anthropologists are presumably aware that when one first encounters a new cultural group, given their concerns and theory biases, often the *differences* will stand out, rather than, or even selectively enhanced over, the similarities between "them" and "us." In the extreme, this factor probably leads to the accusation that much anthropology ends in cultural relativism. The same can and does happen in the contemporary realm of gender studies. Beginning with enhanced differences, and escalated to the extremes of universal patriarchy claims, I hold that there is often a gender Mauss illusion that troubles much of the literature. In what follows, I shall partially revert to a kind of parody of the current arguments that revolve around such notions as "the gaze," "the phallus," and more generally, "binarity." I am implicitly arguing that much of the current debate seems caught in its own myths, which prevent more subtle recognitions of convergences, similarities, and degrees of distinction that occur within the gender wars. The debates often accent divergences, rather than convergences. If the result of the movement to democratically pluralize the gender situation were to be the substitution of simply another myth for the previous one concerning the roles of genders, then only the polarities would have shifted.

"Gazes," Male and Female

One can easily find, within the autobiographical tales above, that the dynamics of idealized normativities may be found both in male and female childhood experiences. If there is some idealized breast size combined with weight limitations for adolescent women, this is matched, for young boys, by the Charles Atlas body size. Anorexia on one side is matched by the search for proteins, or worse, steroids, on the other. This dynamic is a reciprocal one, at base similar but taken in the binary directions that are also reciprocal—at least in much of the current debate.

I have been trying to show that boys—like girls—also have intense insecurities over body image, but more, bodily capacities, even if these take variant forms. The intellectual danger is to extrapolate or exaggerate these differences. I once read a sociological study (and I would footnote it if I were able to find it) in which polls regarding an ideal breast size were taken of both women and men, with the following results: the males reported a much wider range of desirable or acceptable breast size than did the women, who noted a much more precise and narrower range of what counted as beautiful. Does this indicate a keener awareness of "a male gaze" or of a more intense intra-feminine sedimentation of such idealizations than in the case of males? Unfortunately, the poll did not reverse the issue and deal with any phenomenon that might be more acute for males. (In class, I have sometimes reversed the issue by talking about middle-aged male bellies, although it is hard to tell whether it is the affected male or his spouse who is more concerned. Either way, if potbellies are negative signs, then this might imply a negative "female gaze.") The contemporary passion for exercise, body development, or retention of a youthful body, in everything from jogging to the health club phenomenon, seems to be gender balanced in at least reciprocal ways. Beefcake and male "go-go" has as its equally still minoritarian counterpart the female versions of Charles Atlas in magazines showing overmuscled women lifting weights. In these cases, the older binary reciprocities converge and "gazes" may be either male or female.

From whence "the gaze"? Within the phenomenological-related tradition (for the moment ignoring the powerful impact of the neo-Freudian influence of Lacan), a form of the gaze arises with Foucault. In *The Birth of the Clinic* the medical gaze fixes the patient; in *The Order of Things* the objectivistic mode of sight in the Classical era is itself the invention of perception; in *Discipline and Punish* penal punishment and the disciplinary gaze of the Panopticon is enacted on the body of the condemned.[7] All describe a form of "the gaze" that simultaneously objectifies and enacts control over the selectively passive bodies Foucault has a penchant for depicting. This, as in so much of Foucault, is a virtual binary opposite of the Merleau-Pontean active *corps vécu*. Merleau-Ponty's body is the active body, filled with actional experience, in contrast to the culturally fixed and acted upon body of Foucault. Even older than Merleau-Ponty or Foucault is the "gaze" of Sartre's *Being and Nothingness*, in which the primordial conflict of seeing/being seen by the objectifying/objectified gaze is at the root of all interpersonal social relations. In short, the tradition of interpreting sight as an objectifying gaze is already well established in the midcentury of these traditions. Yet all these variants that elevate the objectifying gaze have counterparts in forms of vision that are neither objectifying nor controlling. Buber's *I and Thou* is a relation of mutual recognition and respect of a sacred sight; the objectification of an I-It relation is balanced by the nonobjectifying I-Thou; Marcel's *Being and Having* emphasizes the mutually participatory; and Merleau-Ponty, in his argument with Sartre, says that conflict is only possible upon the basis of a deeper recognition of similarity and participation (*Phenomenology of Perception*). In short, even within this family of philosophical traditions, the notion of the objectifying gaze occupies only one side of a debate. Do the gender wars adapt, post facto, only the one emphasis?[8]

A deep phenomenology of vision shows that such perceptions are multistable rather than simply objectifying as a necessary outcome of vision itself. Only a shallow enculturalization of both a reduction *to* the visual as the dominant epistemological perceptual dimension, and a reduction *of* vision to objectification could support notions of the "gazes" that too often dominate the literature presently.

There is a second objection to the reification of "the gaze." If vision can be both objectifying and participatory in different stabilities, it is also *doubled* by virtue of being both feminine and masculine in overlappings. The gender "system" includes both gazers with the gazes not ever being singularly one gender. Young recognizes that such phenomena as pleasure over dressing, makeup, fashion, and even self-pleasure in one's body (see "Breasted Being" and "Pregnant Subjectivity")[9] is not some simple response to a male gaze. It includes both presumed male gazes and feminine self-gazes, a doubled glance.

The same applies to males. At the age that I have chosen as the target for male coming-of-age with respect to bodies, boys are not yet very, if at all, interested in girls (and vice versa). Preadolescence is largely a time of youthful male bonding, and much of the dynamic, even the negative dynamic of lacking a sports body, is an intra-male phenomenon—but not entirely. The Charles Atlas skinny-to-muscled image is both intra-male, in the sense that it suggests the desire to have bodily power, and aimed at attracting female attention. The gender "system" enters here at least as background. As the transition to late preadolescence occurs, the attentions of the girls in the class intensify even in the awareness of male/female competetiveness and interchange.

First attentions may be ironically negative. In Mark's class, in a phenomenon quite different from anything I remember from the same age, two practices stand out. First, there is anonymous note passing of a varied sort. But one form of such note passing is the "geek tease" note. Whoever is on the bottom of the geek pole receives notes from girls who claim to "love" him. This is obviously a rather nasty joke that the geek may take seriously until his more savvy advisers disabuse him of it. (Mark prides himself in this revelatory role since he is not, in sixth grade, the bottom geek.) The second is the anonymous telephone call. The unidentified caller is insulting, directly and without much nuance. Parents have sometimes been able to identify the caller using caller ID or return-call technocapacities—and sometimes it turns out to be precisely the girl who is the object of some initial attention. In one case, the caller was, in fact, a very popular girl, and she—along with at least one other accomplice—had called many of the boys in the class with similar put-downs. This is all part of the gender system that includes both boys and girls. The girls, not only the intra-male bonders, help establish the set of relations that also usually reflect the hierarchies of who is favored and who is not. The anti-geek tone, later matched by the pro-jock admirations, repeat precisely the social structures of my five-decade precedence over my son's experiences, albeit in differences appropriate to the modes of boy-girl communications that have changed and are relative to the historical differences.

The Phallus

Here I must disagree, by way of an inversion, with Susan Bordo. She claims, "The phallus is haunted by the penis, and the penis is most definitely *not* one. . . . Rather than exhibiting constancy of form, it is perhaps the most visibly mutable of bodily parts. . . . far from maintaining a steady will and purpose, it is mercurial, temperamental, unpredictable."[10] My disagreement is that it is, rather, the penis that is haunted by the phallus. In the perspective of a Merleau-Pontean "I am my body," it is the penis that must somehow live up to the phallic cultural expectations. And it is in this phenomenon that the insecurities of growing up—but also of aging—occur.

While I am certain that I have a penis, I cannot remember when I first became aware of "the phallus" (I suspect it was when I learned something about ancient Greek initiation rituals), and I remain unsure of whether or not I "have" a phallus or of what relation I have to "the phallic." The problem here is that the phallus, in its most recent incarnation, seems to be an invention of the gender wars. This recent version of the phallus seems most usually to entail power, particularly of a political-cultural sort, plus a sense of deep association with masculinity. (It rarely, today, invokes the much more ancient sense of fertility that it had in ancient cultures.) In short, it is another coded variation upon generalized patriarchy. Here a subtle aspect of the current debate frame appears: within gender discussions it is frequently claimed that the unawareness of the phallic is what is to be expected of the male, precisely because the exercise of phallic power is what has been thoroughly culturally internalized. Here again, I disagree by virtue of inversion with Bordo. In her claim that we appear to ourselves not as men but as generic Man, the painfulness of the childhood experience that rests within the two senses of body I have elicited is not one of either the generic Man or of the mythological phallus. Epistemologically, I am disturbed by the forms of unawareness taken over into an unconscious—this because there are so many variants upon any form of unconscious knowledge (an oxymoron). In the deepest sense, I may be unaware of how my neurons are firing, or what capillaries engorge with blood as I experience the beginnings of an erection, but I am definitely not unaware of what is happening in this movement within my experienced embodiment. The physiological unconscious is simply not the same thing as either a psychological or cultural unconscious that is never totally unconscious. Such hybridizations are what I consider to be the deep fallacies of the Freudian-inspired traditions. Physiologically, neurons and capillaries are not experienced precisely because the constitution of such entities is necessarily a constitution of things out there, available only to the observer's perspective, and couched in third-person language. I do not and cannot "see" my brain except as an "other," nor, for that matter, any of the interiorities of the physiological. One does not experience, or even see, the firings and expansions of one's own neurons or capillaries unless imaged in a third-person instrumental situation. The hybrids—which are often inventions of the Freudian traditions— that fill presumably unconscious experiences are not constituted in the same way. This applies to notions of phallic power. As a longtime experiencer of "power" in academic settings, as a chair, a dean, a program director, I was quite explicitly aware of the power (and the risks, ambiguities, responsibilities, and contexts) that I was charged with. I was less aware of its "phallic" nature, except when that, too, became more explicit. Sometimes it did become explicit. Once, at a farewell dinner for a provost, the deans (all male) began a clearly crude "penis-waving" contest about how each had been more aggressive and hard-minded than the others in decision making, and they wondered if the expected in-coming female president would have enough "balls" to carry the day. (She did—but this is clearly an indication that the phallus is not simply male.) This power, which can and does get experienced and which floats in and out of the sign of the phallus, is very different from the nonexperience of neurons and capillaries. It is a hybrid phenomenon that is presumably both out there but experienceable in first-person experience.

Part of the problem here is oversimplification and overgeneralization: there are styles and varieties of power, not all of them phallic; there are phallic implications that are not always related to power; and there are female users of phallic power just as there are male users. The "hard" woman is a cultural icon: Bette Davis movie characters, *Smilla's Sense of Snow,* the genre of female warriors in movies and television including Xena, all complement the already mentioned female bodybuilders.

The phallus is a mythical construct, which like all myths does contain a kind of imaginary power, but an ambiguous power that produces multiple cults and offshoots rather than a singular empire.

It is a part of the malleable and the reconfigurational that, even if more dominant in certain configurations than others, nevertheless spills over any clear and clean boundaries.

Binaries

It should be anticipated by now, given the trajectory of this article, that I believe that unqualified binaries are for simpletons. If there are both male and female "gazes" that overlap, and if "phallic power" can be exercised by both males and females, even if these phenomena are often more strongly associated with dominant, rather than recessive, cultural practices, then any overgeneralization becomes suspect. At the very least, a beginning qualification is one that must insist that contrary to "males are hard/females are soft," some males are soft and some females are hard.

What I am after can be analogously applied and extended from the deep problems with which any philosophical aesthetic must deal: much traditional philosophical work in aesthetics is done by way of attempting to define or circumscribe what can count as an "object" of art. Yet as soon as some definition is set forth, no matter how ingeniously or carefully, in the context of modern or postmodern art, the artists—in reaction, which is part of the avant-garde tradition itself—can and usually do immediately attempt to create an art object that does not and cannot fit the definition. This spilling over has been part of the art trajectory since modernity. In short, art practice overtly revolts against aesthetic proscription as part of the art practice itself. I am contending that, in a similar late modern or postmodern setting, the same dynamics can and do often apply to the "gender wars." The parallelism between deconstructive strategies, which first turn everything into a text and then move from a commonsense notion of texts into increasing emphasis upon marginalia—notes, postcards, margins, intertexts—also applies to the gender discussions. The postmodern result is thus bound to be an ambiguation of *all* boundaries, and indetermination of texts is equivalent to the indetermination of sexes and genders. One can easily recognize this ambiguation in the current gender discussions that, far from binarism, now contain the range from male/female through the ambiguations of queer theory, bisexuality, transsexuality (see Butler), cyborg theory, hybrids (see

Haraway), and onward. One way of defeating overly simple binarism is to ambiguate through indeterminacy.

I am not deploring this development—to the contrary, the ambiguation serves to resituate what can be taken as a problem or set of problems. The current arguments between essentialists and nonessentialists seem to me to be a dead-end set of arguments. If there is anything essentially, i.e., deterministic, about being male in contrast to being female, then any actions that are or can be taken will not threaten that essential difference since it couldn't be exceeded actionally anyway. A hard determinism would be the physiological equivalent to the law of gravity, within or under which all motion occurs. On the other hand, the biotechnological fantasies that imaginatively extend to male "pregnancy" in one extreme or to a totally female-virginal mode of reproduction on the other, but which take increasingly ambiguated concreteness in fertility experiments, could presumably make for unthought-of indeterminacies even at the biological level. Thus if there were some sort of physiological essentialism in today's biotechnological world—at least as fantasized—even such determinacies potentially dissolve. Ironically, nonessentialism ultimately dissolves binarism as well.

This, in turn, points to the now thoroughly culturalized problem as to why there remains such persistence and apparent cultural resistance to gender model change. It makes the problem of how, in spite of radically different contexts, my son's and my own experiences of growing up male remain so recalcitrantly similar. Whatever social construction is, it is not something that appears to be easily made malleable or ambiguous even in the rapidly changing world of technological civilization. In this case essentialism turns out to be thought of as cultural persistence, but effective for all that. In short, there are essentialist and nonessentialist functional equivalents within both the biologically determinable and the socially constructed versions of the discussion. "Conservatives" are those who hold to some persistence of biology, but this essentialism is effectively under attack through biotechnological (and thus through technoscientific-cum-cultural) means, while "liberals," who hold everything is simply socially constructed, come up against the virtually impenetrable persistence of social patterns (whereby boys, denied toy guns by parents determined to redefine juvenile gender patterns, end up inventing their own by turning sticks into weapons, etc.). Functionally, there are both biological and social patterns resistant to change, but equally, such resistances reside within larger gestalts whose interlinking parts make piecemeal changes difficult. In short, I am arguing that the essentialist/nonessentialist debate resolves into the same set of functional problems that are complexly both political and developmental.

Where, Then, Are Bodies Revealed?

After following one person's development in the light of gender experience in the context of embodiment, accompanied by the backdrop of some history of gender debate literature, where does this leave us? My line of argument has attempted to reveal something about bodily awareness in a male context, an awareness that hints that perhaps it is too much a part of the current gender discussion that encourages a Mauss illusion of too binary a situation. But, not willing to speak for either particular or generic feminine experience, what can be said of possible reciprocities of experience? One can turn quasi-autobiographical, which has the advantage of concreteness and particularity. Susan Bordo has become expert in this voice style. One glimpses, within the philosophical context she holds, much that is autobiographical and experiential, with a richness of suggestion the favored anonymous style of much past philosophy has lacked. (Exceptions are also obvious in writers such as Kierkegaard and Nietzsche, for example.) This version of gender discourse, of course, veers toward the literary.

And it, too, has problems. "Reading the Male Body" only shows glimpses of the autobiographical. Bordo obviously knows something about actual male penises and sympathetically contrasts them with the mythical solidity of the mythical phallus. But in a much more autobiographical voice, her piece on "My Father the Feminist," part of the literary turn shows its problem. In this paper, Bordo elicits a rich sense of growing up Jewish in a patriarchal family. Her father dominates, plays the male, and for a long time fails to "get it" with respect to his feminist daughter. The contestative but loving relationship between father and daughter conjures the rich sense of family life in this ethnic tradition. Yet in the telling, literary devices emerge that possibly belie more mundane actualities: Bordo becomes, as in the Isaac Bashevis Singer tale *Yentl the Yeshiva Boy*, the female replacement for the missing (rejected) male boy-scholar. In the arguments that sharpen her wit and intelligence, her father unknowingly begins to provoke the latent feminism that later matures. As in so much good literature, an irony begins to dominate, and by the end of the story, one can wonder—and since I accept the intentional fallacy as fallacy, cannot speculate about the author's intention—whether or not the best way to produce a brilliant feminist might not be to be a patriarchal father. Here, a more complex binarism serves an ironic end but at the same time could be taken as a justification for itself. Thus, if philosophy can produce its own fly bottle, so can autobiography. Similar ironic twists bedevil both the Nietzschean and Kierkegaardian traditions.

How can one combine the richness of autobiographical concreteness with the wider implications of the general patterns, even universality (within the one remaining human species, *Homo sapiens sapiens*)? My own partial answer is that all structures and patterns—and there are these—display *multistable sets of limited possibilities*. This is clearly a phenomenological notion derived from variational theory, but set in a nonfoundationalist context. Gender(s) are multistable. Anthropology surely has taught us this, since even the *National Geographic* has long displayed cultural patterns that vary and invert what is perceived as masculine and feminine in different groups. Certain African males use complex makeup, featuring contrasts for eyes and dark lipsticks, and do dances to attract females—but this is considered masculine in that context. Contrastingly, in the over thirty years I have spent summers in Vermont, the persistence of lipstick-free, straight long haired, jeans- and flannel-shirt-dressed women, bespeaking a certain country air, remains in place and in contrast to today's black or black-outlined lip designs of Stony Brook undergraduate women, yet both can easily be identified with a kind of feminine statement within the multiplicities of American culture. As for contemporary masculine, the outrightly boring universality of the baseball cap, bill forward, backward, sideways, is part of that statement, today accompanied by multiple earrings in one or both ears, the latter body fashion unthinkable in my boyhood.

Late modern multistability is itself part of the pluralizing of cultures—a bricolage of culture fragments adapted easily by both males and females in late industrial culture. One does not know what this will do for patriarchy, for feminism, or for the gender discussions, but one can know that perhaps the myths of the past today appear quite ambiguous in the light of much actual practice. And all of us are, for better or worse, embedded in these more overtly recognized multistabilities. Yet for all this, the horoscope calendar displayed in our Chinese takeout in our village, which displays two very young children, clothed only in underpants, each opening up to do a self-look, expresses the obvious: differences remain.

Epilogue

In fairness, I need to bring the readers up-to-date concerning Mark. This essay was written when Mark was still in sixth grade; he is now in the ninth grade in junior high school. And while there remain echoes of the elementary school tensions, much of the earlier situation has been modulated. First, the new school situation is much larger and more diverse, with new sets of friends and groups, thus only some of the previous set remain core friends. Second, the expected growth has begun with a vengeance, and he is now well situated among the average-sized boys, exceeds his mother's height, and is rapidly approaching mine. But third, he has exceeded all expectations with respect to his alternative activities. In the last month of elementary school he entered and won a major prize for piano composition that was publicized throughout the region and in the schools. From this, he went on to perform the composition in various university and conference settings within a new group, the small number of prize-winning composers. The result was a kind of confidence that today has spilled over into writing different styles of composition (electronic plus piano) and continued high-level performance in class. In short, he has partially transcended some of the in-group values expressed above and begun to find his own métier. This fall he accompanied me to Vienna and shared a presentation on "Moog, MIDI, and More: Electronic Music" with Trevor Pinch and me, his first academic conference presentation and part of the new mode of father-son identification. (If this sounds too much like the ravings of a proud father, you will simply have to take that into account.)

Bodies in the Philosophy of Science

Part II

Chapter 3
Visualism in Science

It will be the contention of this chapter that one of the cultural habits of the sciences is to produce, display, and reiterate what counts for evidence in visual form. I call this science's *visualism*. This cultural habit has been accelerated in late modernity through the sophisticated development of imaging technologies, which now transform ranges of phenomena that include, but also exceed, all human perceptual capacities and *translate* these phenomena into visual forms. This constitutes a unique form of technological constructionism and is thus relevant to this context. These techniques, I contend, constitute a visualist hermeneutic that has become the favored mode of evidence display within most sciences.[1]

By characterizing science's preference for visualism as a cultural habit, there is an implicit claim here that this cultural "choice" is not a necessary one, but a perceptually and culturally contingent one. I shall try to show how this is the case through three strategies. One strategy is to take some account of what I shall term whole-body perception by means of both phenomenological and nonphenomenological accounts of perception. The second strategy is to make selective reference to incidents in the history of science as it increasingly condenses its evidence presentations into visual forms. The third strategy is to locate much of the production of visual displays in science's technologies or instruments and to show how the various information-gathering devices are increasingly developed to make just such visual displays.

Whole-Body Perception

I begin by drawing from phenomenological traditions that explicate perceptual phenomena through bodily actions. Phenomenologists from Maurice Merleau-Ponty through Hubert Dreyfus claim that human bodily action is the focal and necessary basis for human embodied intelligence.[2]

In actual experience there is a constancy of what I call whole-body perception, in the sense that our perceptions occur as a plenary gestalt in relation to an experienced environment. We interact with the world around us. It is also the claim here that our whole-body perceptions are sensorily synthesized in our interactions with a "world." Unlike the older traditions of discrete and separable senses, phenomenology holds that I never have a simple or isolated visual experience. My experience of some object that is seen is simultaneously and constantly also an experience that is structured by all the senses. It takes some deliberate constructive manipulation or device to produce the illusory abstraction that could be called vision by itself. Thus, as in empirical psychological experiments, when various color illusions or gestalt deformations are inspected, the experiment deliberately designs the situation to dampen and displace the ordinary whole-body engagement with surroundings. To look through a tube or to be seated in a darkened and silenced environment, in effect, produces a quasi-illusion of seeing itself or an abstract seeing.

There is here, too, a doubled phenomenon. The framing of a situation so that only its visual dimension is accounted for changes precisely that situation that shapes attention or focal action itself into special forms. Whole-body, ordinary activities occur within complex environments in which much is simultaneously occurring. This is the perceptual field that constitutes our immediate environment, and it is never simple. We therefore must focus our attentions upon those aspects of this complex field in such a way that our intended actions may be carried out. Thus, if I am a bird-watcher, I will focus upon the goshawk chasing the kingfishers over my pond, rather than attending to the music coming from the floor below (although the focus cannot eliminate the sound of the music, it pushes it into the background). Similarly, although in a more controlled form, the instructions that determine what it is that I am to look at or see in the constructed experiment rely upon this ability to perceptually focus. In this case the very ability to focus helps to enhance the quasi-illusion of a pure visual phenomenon by subduing the other sensory dimensions.

Overall, this analysis aims to show that regardless of deformations or manipulations, whole-body or multisensory dimensional perceptions remain constant in environmentally interactive situations. The focal attentions that concentrate upon visual aspects of phenomena are nevertheless situated within the more primary synthesized whole-body perceivings.

One can see how this plenary or multidimensional perceptual action can and often does come into play with some examples of scientific practice that entail whole-body perceptual actions related to concrete scientific work. Imagine a physical anthropologist searching the ridge of some African plain. There is simultaneously the bodily motility of the anthropologist as she walks along the ridge engaging her active visual scanning, focally searching the field for some small indication of a human fossil. There! It looks like a tooth or some teeth amid the stones and even possible nonhuman animal remains littering the surface. The anthropologist kneels for a closer look and then to feel and probe into the dirt—the tooth protruding is part of a jawbone, so with care, perhaps now with a tool, the probing begins with fully engaged bodily-plus-tool actions at the site. In this illustration, it is *ordinary* full-body engagement that characterizes the discovery situation. Informed vision is employed along with tactility, bodily motility, and the rest of the interaction with the immediate environmental world of this situation. The initial point here is that the scientific activity is occurring within precisely this full-body perceivability of the action of the anthropologist and is not exclusively or even necessarily primarily visual in character. The tooth-jawbone is seen, felt, handled by the anthropologist, and, except for the highly trained and critical perception involved, no different than in cases of ordinary objects.

If we now turn to a different science practice, for example, contemporary astronomy, a somewhat different pattern emerges. Here the scientific object must be experienced very differently since it does not have any direct bodily presence at all. For instance, a recent *Science* report indicates that earlier Hubble Deep Field (HDF) exposures (yielding images) seemed to suggest that as the telescopic probe got deeper into space-time, indicators of star formation seemed to slow down. A new report, however, showed intense gamma ray blasts that suggest star formation activity as far back as detectable (thirteen billion light-years), thus contradicting the Hubble observations.[3] While both instrument-sets yield visualizable images as part of the evidence, the visual displays are the results of instrumentally constructed or mediated processes that are *translations* into visual forms of phenomena themselves not directly available to human perception (gamma rays, or in the case of HDF, indications of an empty field.) In short, the whole-body perceivability situations are quite different in these two sciences. Yet in both instances visual displays continue to become part of the eventually developed evidence for claims being made.

Perceptual Variants

What I have so far emphasized is the way in which visualization lies embedded in scientific practices and is an enhancement of certain dimensions of whole-body perception. Here, I wish to shift ground a bit to deal with what I take to be a traditional prejudice concerning a presumed intrinsic perceptual superiority of vision over any other human sense, thus implicitly justifying science's cultural choice as in some way natural. I do not want to enter this debate deeply, but as a longtime researcher into auditory phenomena I can at least indicate that this sensory dimension has some counterpart claim to discriminatory equivalence. At the physiological level, there is no distinctive difference between seeing and hearing with respect to detecting duration (both at 0.01 seconds), but the sense of hearing actually carries longer before nerve damping occurs with respect to a single stimulus. The discriminations trained musicians can make are as fine to finer than those for trained sight, and there is no counterpart for "perfect pitch" in sight such as there is for hearing.[4] The range of auditory perception with respect to the continuum of wave phenomena (for humans from 20 to 20,000 hertz) is vaster than for seeing within the optical range of light waves.[5] Yet another indicator of auditory equivalence may be found in recent experiments that have created auditory equivalents to inverting glasses. Dutch researchers have recently fashioned distorting "ears" that deflected sounds from a speaker on a robotic arm such that locations were inverted (above/below; right/left), yet, in an exact parallel with inverting glasses, the "ear wearers had compensated and were able to make correct judgments. The surprise came when they finally removed the ears: they were immediately able to adjust to their old ears. 'That's quite astounding,' says Fred Wightman, psychologist at the University of Wisconsin, 'it may mean that locating objects by sound is a more complicated cognitive function than with vision.'"[6] Other indicators come from ethnography, which locates cultures other than ours in which auditory perceptual dimensions play very strong roles, such as those in deep rain forests where sound plays a stronger lifeworld role than in our visualist culture.[7]

Turning from perceptual variations to instrumentally mediated perceptions, it may be noted in passing that scientific instrumentation has not itself always been a matter of visualization either. Bettyann Kevles's study of medical imaging, *Naked to the Bone*, indicates that earlier in the century, after the introduction of the phonograph, which captured transient sensual experience, Edward Bellamy suggested that "hearing is the sense of the future, and coming none too soon to rescue eyesight, which 'was indeed terribly overburdened previous to the introduction of the phonography, and now that the sense of hearing is beginning to assume its proper share of the work, it would be strange if an improvement in the condition of people's eyes were not noticeable.'" She goes on to point out that medical technologies were also often more auditory than visual—the stethoscope, percussion of the chest, and even recorded phonographic sounds.[8] Similarly, sonic transponders and other early submarine instruments often were "read" auditorily prior to visual graphing translations, which were developed later. Of course, this brief hope for auditory modeling preceded the invention of motion pictures; later, motion pictures with sound and listening gear were replaced by graphic radar-like screens for transponders.

Science's Visualist History

To this point I have tried to show that visual displays, at least as contrasted to either whole-body perceptions or to possible auditory ones, had to have been something of a choice. But this is not to say that such a choice is individual—it is more a historical-cultural event.

If we take Leonardo da Vinci (ca. 1500) as our first individual moment of this early modern shift to visualism, we see a double transformation of how visualization occurs: a shift to vision and its reduction to a certain kind of vision. The shift to the visual is an enhancement of the visual over and above and often to the detriment of either full perceptual or nonvisual perceptions. For example, descriptive anatomy at the time was often in tactile and olfactory terms that referred to how an organ felt (hard, soft, pliant, etc.) or smelled (putrid, metallic, etc.). Da Vinci reduces this anatomy to a structural and analytical set of drawings, which visually depict tendons, muscles, and veins (later followed by Vesalius's famous anatomical studies, ca. 1540).

Early modern visualism was also technologized. It is well known that one of the favorite visual "toys" of the Renaissance was the camera obscura. What often goes unremarked was the highly important role this optical instrument played in the development of so-called Renaissance perspective. Alberti (ca. 1437) apparently used the camera obscura quite regularly. He may have been among the first to draw by the lines.[9] Note that the camera obscura reduces three-dimensional objects to two-dimensional images. That is, this isomorphic "reduction" is an artifact of an early imaging technology. Da Vinci, in turn, was the first Renaissance describer of a detailed camera obscura (ca. 1531).[10]

The second moment in this impressionistic history is Galileo Galilei. But the Galileo of this history is the visualizing Galileo, the Galileo of the telescope (ca. 1610). A maker of many optical instruments before his discovery of the telescope through Paolo Sarpi (ca. 1609), he had already been the maker of surveying instruments and later used the microscope.[11] It was, however, through the results of his telescopes, of which he made approximately a hundred, that his form of science's visualization entered this trajectory. Although it apparently took some time from his first telescopic results before he turned the new instrument to the skies, once he did he was, within three months, publishing his results in *Sidereal Nuncius*. He was immediately able to formulate a simple instrumental realism by recognizing that the new visual phenomena he was seeing were real. His observations, which have stood the test of time, include the surface features of the Moon, sunspots, the four main satellites of Jupiter, and the phases of Venus. Not only were these heretofore unseen phenomena, but the new instrumental mediation made possible the transformation of perceived space-time (for example, reduction of apparent distance and magnification of apparent motion) not possible before the telescope.

The third moment of this simplified cultural history arrives with the rapid acceptance of photography. This imaging technology was, in fact, one of the most rapidly accepted and adapted imaging processes in the history of science. Daguerre's process was publicized in 1840 by a photograph of the gibbous Moon; by 1842, photographs were being made through spectrographs, micrographs, and other compound imaging technologies. With faster exposure and shutter speeds, time-stop photography began to depict motion studies of interest to nineteenth-century science. But why was photography such a breakthrough? The optical technologies that magnified phenomena beyond the limits of ordinary vision yielded new and unexpected entities, but to show or demonstrate these one had to have each observer take his or her own sighting or, more commonly, make a drawing of the object. This meant that two matrices for subjectivity had to be surpassed, that of skilled vision through the instrument and skilled reproduction by drawn representation. Similarly, Leeuwenhoek was the first to report the microscopic sighting of spermatozoa, but again the problem was one of a drawn representation. Photography, not unlike the camera obscura, automatically reduced the object to an isomorphic and realistic fixed image on the photographic plate. And it accomplished this without subjectivity, as it were, by means of a mechanical process. (Here I deliberately add the nineteenth-century concern for subjectivity, which was not a topic of discussion earlier.) Technologies, traditionally taken as nonhuman, thus serve to standardize perceptual results in imaging processes.

Before leaving photography, one more facet to this process that makes it scientifically interesting needs note. Although the early processes called for long exposure times, limiting imaging to stable objects, more and more rapid exposure times allowed later photographs to manipulate or transform time. Muybridge's 1878 photographs of animal and human locomotion were used as evidence for previously unknown dynamic phenomena; the Mach brothers showed shock waves for the first time; and much later strobe techniques "stopped time" to reveal even more microfeatures of dynamic phenomena. This accumulated success of photographic imaging is noted by Kevles: "By the 1890's photographs had become the standard recorders of objective scientific truth."[12]

I now temporarily break off from this narrative history of scientific visualism. I have quite deliberately emphasized not only the turn to visual phenomena (including their development in imaging technologies), but also their largely modern features, which include realistic isomorphism, representability in images, and analogues to ordinary visual phenomena.

I also want to note that all imaging technologies, from the beginning, transform perceptibility. Even the simple visual transformations of the camera obscura are significant (reduction from three to two dimensions, inversion of the image in relation to the referent object, etc.), while those of the earliest magnification technologies (microscope and telescope) took human vision beyond its ordinary bodily limits—at the same time retaining the obvious analog qualities of ordinary vision (animalcules within rainwater or mountains on the Moon) and making the sometimes strange phenomena nevertheless recognizable. It was initially this mechanically reproduced isomorphism that gave imaging its scientific advantages. As early as 1888, P. J. C. Janssen claimed:

> The sensitive photographic film is the true retina of the scientist . . . for it possesses all the properties which Science could want: it faithfully preserves the images which depict themselves upon it, and reproduces and multiplies them indefinitely upon request; in the radiative spectrum it covers a range more than double that which the eye can perceive and soon perhaps will cover it all; finally, it takes advantage of that admirable property which allows the accumulation of events, and whereas our retina erases all impressions more than a tenth of a second old, the photographic retina preserves them and accumulates them over a practically limitless time."[13]

Structures of Vision

The primary model of visual perception I have been following is derived from a phenomenology that holds to the primacy of an actional body/environment relativity. Many of the insights of gestalt psychology are also related to the phenomenology of perception (Koffka and Kohler were students of Husserl). While I cannot be thorough here, some salient features help underline the rationale for science's visualism.

Both gestalt and phenomenological psychologies emphasize the importance of the variable figure/ground relationship. Figures or focal objects only appear within and against a ground, but figures are variable within any given ground. Continuing the use of standard scientific examples, the ability to pick out a comet or a supernova, either from observation or from alternating photographic images, utilizes this visual skill.

Full instantaneous gestalt pattern recognition is another visual structural skill that comes into play. A personal example here shows the power of this form of visualization. My oldest son and his wife recently gave birth to their first child—quickly, in the fashion of high-tech households, they e-mailed a digital photo to me. My computer, however, had text-only capacity and thus some twenty pages of gobbledygook came forth. One simply could not make out what the referent object was from the data itself. When sent to my wife's better machine, the picture emerged on a single page with the aforesaid instant recognition. A picture is worth far more than a thousand words these days.

Assuming the full panoply of contemporary imaging previously noted, reiterability is yet another valuable feature of scientific visualism. One can return, again and again, to the image to detect features overlooked or previously unnoted. It is the instant and reiterable feature of the visualization that makes it valuable for science.

Pattern shifts or repetitions themselves are part of the recognition process. In a parallel to the data/photo example above, the scaled repetition patterns of such imaging as fractal sets immediately display patterns that would be either quite indiscernible in the data or misunderstood.

To this point I have held to a parallelism between full, actional *ordinary* perceptions and their scientifically guided, but merely *enhanced* perceptions. For that reason, I have restricted myself to analog visual phenomena. A scientific perception within this limitation may be a more informed perception, may be more skilled in various focal concentrations, and may hermeneutically "see more" within any given object seen, but it is a perception that still retains its gestalt and patterned qualities. Galileo, upon seeing the four satellites of Jupiter, virtually instantaneously (the "aha phenomenon") recognized that here was another body in the heavens that had "planets," as he called them, circling it. At the very least, the Earth was not the only such body with satellites. Yet there was simultaneously a much bigger transformation of what would count for scientific vision implicit in these early observations.

Second Sight

The transformation of vision, of which I am hinting, is what happens to vision through technologies of vision, or instruments. I have suggested that, at least since early modernity, the role of instruments in science has been crucial—but this is also a *perceptually crucial* factor in the constitution of science's visualism. If the camera obscura geometrically rearranged visual objects—and it belongs to the earliest movements of the geometrical method, explicitly playing a role in both Descartes and Locke[14]—lensing opened the way to the much more radical transformations that were to follow. Lensing, in the telescope and microscope, transformed the phenomenological sense of space-time. Apparent distance, in technical language today, is a description of the changed relative distance between observer/observed. It is relativistically equivalent to say that the telescope brought the mountains of the Moon "closer" to Galileo or Galileo "closer" to the mountains. It was in the difference of direct and telescopically mediated distances that the sense of space was transformed. Similarly, the microscope repeats for the miniscule the same feat of transformed sense of distance, and phenomenologically speaking, it may be said to be equivalent regarding telescope and microscope. (Moon mountains and paramecia both take up focal places within the now instrument-mediated apparent space and are "same-sized" objects within central vision.) These transformations, partially noted by Galileo and his detractors, were either accepted or rejected precisely because of the transformational changes to vision. Eventually, of course, the transformations won out, and science became, in modernity, a process of *instrumental realism*. What could be seen through the lensing systems was taken as real, in part because it retained its analog qualities to unmediated vision.

Magnification, whether micro or macro, already holds the secret of getting beyond vision insofar as it is limited to direct and ordinary embodied vision. But I would contend that it is only by introducing a much more drastic set of variables that what I call "second sight" emerges. Galileo accidentally stumbled upon one such phenomenon in his invention of a heliograph, through which he imaged the first observed sunspots. (A heliograph in this instance was a screen upon which the telescope cast an image, a camera obscura–like device attached to the telescope.) Sunspots, unlike mountains or satellites, entail more than relative spatial distancing. To observe a sunspot with the naked eye is to incur blindness—a different form of embodiment is needed to see a sunspot and that is what the heliograph performs for vision. It not only enhances analog vision, it displaces it with a type of second sight. This second sight becomes a unique, instrumentally constituted scientific object for sight.

Second Sight and Technoconstruction

I want now to follow only in a limited way the trajectory that I have opened up with second sight, or more radically, technologically constituted scientific vision. I have previously noted that, in the very earliest days of scientific photography, compound instruments were being used (for example, the spectrograph in 1842). A spectrograph does not image an ordinary analog visual object (such as a star), rather it depicts the spectrum of light emitted and displays it as a spectrum of colors (which can be analyzed as signs of chemical composition). This deliberate set of manipulations becomes even more typical for late modern science.

Imaging becomes, in the twentieth century, second sight imaging, which employs imaging from the infrared and ultraviolet ranges of the optical spectrum—beyond ordinary visual capacities but instrumentally translated into visible patterns.

Once beyond the optical ranges, the same translation capacities are employed for wave phenomena ranging from gamma to radio waves; again, these are made visible through the translation into visibility.

False color manipulations are used, under different conventions, to visually depict degrees of intensity or wave phenomena well beyond ordinary human bodily capacities to envision. Deliberate enhancement and contrast techniques are employed, particularly in computer tomographic processes, to better display easy-to-miss features of the target object (techniques used from astronomy to medical imaging).

I am terming all of these visualizations a form of second sight that lies beyond ordinary or whole-body engagements, yet remains a visualization. Second sight is a hermeneutic style of envisioning phenomena. It retains all the advantages of gestalt and phenomenological visions, yet it is a translation into the visible of phenomena that lie beyond literal vision.

The Anthropomorphic Invariant and the Latourean Laboratory

In concluding, I want to draw two implications from this narrative of scientific visualism. The first regards what I call the anthropomorphic invariant and relates to the necessity of the bodily perceiving person who does the science involved. I began with a claim that the culture of science "chooses" to carry out its operations within a certain style of perceiving that I am calling visualism. Yet invariant within this cultural style is the necessity for there to be a bodily perceiver. At first, and at bottom, this perceiver is an ordinary and direct perceiver— this is to say, there could be a sort of lifeworld science in the old-fashioned Husserlian sense. But historically, science since its earliest modern forms became more than directly bodily perceiving in its observations. It became technologically embodied through its instruments, although these have frequently been imaging instruments. Yet with, through, and among these instruments, the scientist also always remains a bodily perceiver—that is, the reflexive retroreferent of scientific activity. And this perceiver remains the anthropomorphic invariant throughout the entire spectrum of observation, from direct to translated and technoconstituted imaging, from first sight through second sight.

My second conclusion has to do with the sociocultural construction of science in its current forms. Insofar as it retains the visualist trajectory it took upon itself in early modernity, it now functions very much like what I have called a "Latourean laboratory." By this, I mean that the hermeneutic insight employed by Bruno Latour in *Science in Action* remains operational here as well. Laboratories are, for Latour, places where scientists work and where instruments are employed. But instruments, for Latour, are inscription-making devices that produce visual displays: "I will call an instrument (or inscription device) any set-up, no matter what its size, nature and cost, that provides a visual display of any sort in a scientific text."[15]

The invariant of the bodily perceiver, the anthropomorphic standard, is linked to the complex hermeneutic devices of instruments or inscription-making devices to produce visual displays, some simple and direct, others complex and indirect (translation devices), but all produce the visualizations that count as evidence in late modern science.

Finally, I want to point to one interesting generalization that emerges from this very short examination of the visual and technological constructionism of contemporary science: the better the image and the better the information revealed, the more highly constructed and the more thoroughly technological has been the process of producing this knowledge. If this generalization holds—and I think it does—then we are today a very long way from the simplistic notions of objectivity and mechanical representationalism of the late nineteenth century. Were we to develop an epistemology based upon current science praxis, we would have to include a deep and critical analysis of precisely this instrumental, technological constructionism.

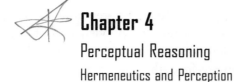

Chapter 4

Perceptual Reasoning
Hermeneutics and Perception

A so-called hermeneutic philosophy of science is, within English-speaking contexts, both a recent and a minoritarian tradition. Its roots are more deeply Euro-American than Anglo-American, and the latter developments remain, even today, dominant in those contexts. Yet today there is a certain strength to be found in a generation of philosophers both appreciative of science and tutored in the Continental European traditions of philosophy.

This is not to say that a European philosophy of science itself became distinctive out of those same Continental traditions. For while Heidegger and Gadamer in the German context, and Bachelard and Merleau-Ponty in the French, had things to say about science and were applied by some to scientific practice, it is hard to point to any organized or unified tradition in a phenomenological-hermeneutic (hereafter P-H) philosophy of science per se. Historically there were at least two reasons why such a tradition remained marginal. First, the relationship between the principals in the P-H tradition and practicing science was frequently perceived as a critical one. Unlike its contestants, the positivists, the P-H philosophers never valorized science nor saw it as the utopian answer within the history of rationality. Contrarily, they tended to see science as both a derivative and limited enterprise in relation to deeper lifeworld, historical, and cultural tendencies. Second, the principals often tended to interpret science, in spite of its ranking as a derivative practice, much more precisely in positivist terms than was warranted by what should have been a more penetrating P-H analysis of science itself. That is to say, what positivism had to say about science as primarily a hypothetical-deductive and largely theoretical enterprise tended to be taken as correct by the P-H philosophers as well. One can perhaps see, retrospectively, that these two tendencies reinforce each other. If science is thought to be more theoretical-conceptual than it is or was, then it clearly is more unlike a praxical, sensory, and bodily immersed lifeworld than is daily life. Thus while there remained more familiarity in Continental Europe with P-H traditions, and perhaps more appreciation, both by virtue of perceived criticism and a too-narrow and too-conceptual view of science by the principals, P-H philosophy of science remained more latent than not.

Since the sixties, however, much has changed. In the Anglo-American world of philosophy of science, the overthrow of positivist philosophy of science has been fairly obvious. From Kuhn, Feyerabend, Popper, and Lakatos on, the gradual reinterpretation of science in much more praxical, historical, and, more recently, sociological terms has proceeded apace. Today philosophers characterizing science no longer depict it as primarily a theoretical-conceptual exercise, a utopian expansion of a unified knowledge, or a value-neutral and exceptional human sociocultural activity. Instead, science is seen as both more pragmatic, finite and limited, and socially-culturally constituted, even up to and including possible deep gender biases and Eurocentric features.

Moreover, there is no such thing as the philosophy of science; there are only many perspectival philosophies of science. This proliferation, from a field occupied by only a small number of philosophers worldwide, remains more than usually productive in terms of books and articles published and cited.

I have taken the trouble to rehearse this very brief sketch of recent philosophy of science directions to locate more precisely what I shall attempt here. I want to focus on the minoritarian P-H traditions within philosophy of science, particularly those that have appeared in English-language contexts, and take account of some special features and particular controversies within this tradition, which nevertheless have ramifications for the wider spectrum of philosophies of science.

Were I to try to isolate three main contributions to contemporary philosophy of science where P-H traditions have made impact I would list the following: (1) the "strong program," (2) the political movement, and (3) hermeneutic philosophers.

One of the strongest traditions in the reinterpretation of science has come from the philosophically minded sociologists sometimes associated with the so-called strong program. In its most radical form, the social constructionists view science as no different in principle than any other social institution or practice and claim—with admittedly very different degrees of radicality—that the products of science are socially constituted. At the least this is to see science as a particular form of social praxis, to understand it as an institution (implicitly as open to and prone to fallibility and values as any other institution). Science no longer is then essentially a theory-concept-producing factory with special privilege within the fields of knowledge.

It may be noted in passing that one of the principals in this movement is Andrew Pickering, whose *Constructing Quarks* is a high-level example of this kind of sociologically oriented P-H analysis.[1] In other words, while not all of the sociology of science is driven by P-H insights, significant borrowings do reemerge in the programs of this quasi-philosophical tradition and today are prominent in the discussion.

Related to the sociological analysis that reinterprets science as an institution is the movement to see institutional science as intrinsically political. Here the movement is even farther away from metaphysical, conceptual, theoretical interpretations and moves into both critical theory (Habermas) and knowledge-power (Foucault) applications to science-as-institution. Joseph Rouse's *Knowledge as Power*[2] is a good example, but also included here is the newly emergent field of feminist critics who have put a particular gender perspective upon the same subtle political dimension of science. Sandra Harding's *The Science Question in Feminism* is a foremost example of this reinterpretation.[3]

The third group of philosophers from P-H traditions to make a contribution are sometimes called the hermeneutic philosophers of science—although I am uncomfortable with that term for reasons that will emerge below. And while this group continues the understanding of science as an institution, as those above, its attention has been more specifically on the epistemological praxes of science itself.

There are at least two broad areas of consensus among these philosophers. In contrast to the classical P-H giants (Heidegger-Gadamer or Merleau-Ponty-Bachelard), the contemporary P-H philosophers deny a strong difference between the lifeworld and the scientific world, although the forms of constraint and specialized objectives to be attained may be differentiated. And the mediation between a lifeworld and a scientific world is to be located particularly in what I call a praxis-perception model of constituted knowledge. The mediation focuses upon a particularly material aspect of science, its technologies or instrumentarium. Put in simplest terms, bodily insertion in an environment in which the interrelation is praxical and perceptual is expanded and modified by a technological instrumentarium. It is through instruments that transformed perceptions occur and new "worlds" emerge, but any new world is itself a modification of life-world processes. Science, in this view, becomes more a product of bodily relativistic perspectives enhanced through a concrete and material instrumentarium.

Nor are the P-H philosophers alone today on this issue. I have argued in *Instrumental Realism* that the Euro-Americans no longer stand alone in this emphasis, but are today joined by other philosophers of science more usually associated with Anglo-American traditions.[4] Thus if Hubert Dreyfus, Robert Crease, and I might be recognized as Euro-American, Ian Hacking and Robert Ackermann—to whom we might add certain aspects of Peter Galison and Bruno Latour—fill out a field of praxis-oriented philosophers who appreciate what I shall call the role of "perceptual reasoning," which takes its shape through an instrumentarium.

Perception in Reason

Regarding science, philosophers interpreting it from whatever persuasion would have to agree in some sense that science is observational; observation entails perception; and perception-observation is often, perhaps always, mediated and constituted instrumentally/experimentally. What philosophers of science might disagree about relates to what direction knowledge takes. Does one first theorize and only later go on to deal with some predicted result in an observational context? Or does some newly observed phenomenon lead to theorizing? Or is there an interaction? They also might disagree about how theory-laden or purely given an observation-perception might be, or even about what counts as perception as contrasted with sensation, judgment, etc. And within the minoritarian P-H traditions, one finds disagreement about how perceptual a hermeneutic process might be as contrasted with how hermeneutic perception itself is. I shall not outline here the variant positions taken by the philosophers mentioned above, but instead turn to much more direct examples of perceptual reasoning in the context of instrument use.

Eyeballs and Instruments

Historically, the sciences that have taken dominant place in the interests of philosophers of science—physics and astronomy—were early related to an optical instrumentarium. Galileo, the chief symbolic figure in early modern science (and in Husserl's interpretation of it as well), was a technologist who developed and perfected some one hundred telescopes in his career, but he was also the physicist-astronomer who first brought to Europe's attention new perceptions mediated through the telescope such as mountains on the Moon, the rings of Saturn, and satellites of Jupiter. But Galileo's visualist science, in which vision was extolled over all other senses, took its position within an overall Renaissance celebration of the visual. Perception as visual correlated with optics as the instrumentarium.

This vision, of course, was not only the dominant form of perception, it was a vision in which eyeballs and instruments interacted in particular ways. Motions, shapes, and measurements were the selective features that claimed center attention. In this respect, one could say that early modern science, in its perceiving, doubly reduced plenary perception was both a reduction to vision and a reduction of vision. It was this forgetfulness of the plenary or whole-body perception that Husserl called the forgetfulness of science in the "Origin of Geometry,"[5] and this forgetfulness led him to claim that the primal and plenary perception of the lifeworld was far from the abstractness of science.

This visualist trajectory, set in motion in early modern science, continues unabated in many of the sciences most related to original physics and astronomy. This is so much the case that many of the instruments in the contemporary instrumentarium could well be called visual translation instruments. For example, interplanetary probes— say of Venus—take instruments that use radar "sound" probes to map the surface through the constant cloud cover. The data are digitally transmitted to the master station, but however constructed and transmitted, the "translation" machine is designed to produce a photograph-like visual display of the surface features that shows rifts, valleys and mountains, and lava flows—a visual display. "To see is [still] to believe" in this setting. Similarly, while early sonar in submarine contexts was both conducted and interpreted auditorily (the observer became trained to detect location and direction by the ping sound and time spans), more recent perfection of the instrument yields, again, a visual display where the target is figured against a topographical ground. In short, the near distance of ordinary vision, where the perceptual "noema" is what is seen on the screen, is presumably the lifeworld reference space for "seeing the world."

Today the range of imaging technologies, in which an ever-expanding set of probe technologies produce a visual display, is indeed most impressive. We are even able now to image an atom (and it is probably highly indicative of the state of corporate technoscience that the first published image of a set of atoms was produced by IBM with the company logo "IBM" spelled out in manipulated atoms). I shall return to some of these.

Here, however, I want to take a different and more specifically phenomenological reexamination of lifeworld instrument processes in order to analyze aspects of what I call perceptual reasoning. Husserl's claim that science distances itself from the lifeworld by "forgetting" the plenary qualities of lifeworld primary bodily perception has a point if and only if one takes scientific self-interpretation as given. If one were to take it so, one might think that early modern science was monosensory in its visualism, whether speaking of ordinary vision or of instrument-mediated or instrumental vision.

To anticipate now one of the two dimensions of the reexamination I am going to undertake, one might also note that even earlier than Galileo the drive to create a visual and thus primarily monosensory science was also highly developed by Leonardo da Vinci. Long before Vesalius developed his explicit anatomy, da Vinci had already taken the task of analytically and descriptively showing interiors of the human body in visual form. His exploded diagram drawings of a fetus, musculature and internal organs, etc., all anticipated later scientific anatomies. Leonardo's engineering vision of the three-dimensional exploded diagram, still strikingly modern, was a universal vision for him and applied equally to corpses and machines. His imaginative (and usually unworkable) technologies of pumps, flying machines, and war machines were, like the fetuses and muscles, stylistically the same as analytical-Euclidian exploded diagrams.

A Partial Phenomenology of Scientific Perception

What I have sketched above is a science that perceives and represents. It does so through a preferred visual form of observation that is often, at least implicitly, taken to be monosensory. Its ideal observer is, moreover, placed in as high or godlike a position as possible and is motionless, the point of view from which any world may be seen. But this is still early modern science and today, even within science, must be taken to be an archaic mode of seeing. To overcome this view, I shall now take two phenomenologically guided forays into the practices of sciences that will show something quite different.

Husserl's critique of the early modern trajectory of Galileo and Descartes contains the observation that such science "forgets" the plenary perceptual and bodily base of the lifeworld. At one level Husserl perhaps is correct—but at another he is wrong and overlooks the way in which instruments as technological embodiments of science function to relate scientific praxis to the lifeworld in all its plenary richness. I shall now try to show both how early modern science's forgetfulness and then later Husserl's forgetfulness may be reinterpreted.

My first foray is quasi-historical. What if the science that would have drawn philosophy of science's primary interest had been medicine instead of physics-astronomy? Interestingly, the practice of medicine, at least insofar as its primary object of interest remained and remains the living body (in a medical sense), has a different constraint system placed upon its investigations. It can intrude upon the living body only with the danger that the intrusion itself may deletriously injure precisely what the doctor wishes to preserve or cure.

This is why, in a particularly telling way, in early modernity the developed Renaissance interest in pathology, anatomy, and, in short, the examination of dead rather than living bodies is instructive. Here the probe or intrusion no longer threatened the integrity of the living body, but the object of knowledge was itself dead and thus in some unknown (and still unknown) way different from its live counterpart.

As an aside, it is interesting to note that most anatomy in da Vinci's time reflected more precisely the practice of the pathologist and precisely in the multisensory modes of plenary perception. Anatomical descriptions included, often predominantly, the olfactory, tactile, kinesthetic dimensions perceived in the examination of the corpse—how it or its parts smelled, felt in terms of hardness or soft-ness or textures, the resistance or lack thereof of the organs—all en-tered into the description. What da Vinci did, and his trajectory was largely followed by Vesalius, was to make prominent, virtually to ex-clusivity, the visual depiction of the body; anatomy becomes visual-izable anatomy in early modern science. Ever more minutely dissected later, the microfeatures of the body were investigated through micro-scopic developments; scientific anatomy followed the already noted visualist trajectory.

But this was never the whole of therapeutic medicine. For exam-ple, were we to jump all the way to the nineteenth century, the exam-ination of a living patient is undertaken by a whole series of hands-on practices: palpation, finger probes (males are aware of the way a prostate exam is conducted), and, in a technologically mediated way, auscultation through the stethoscope, in which the highly trained per-ceiver uses his or her ears to sonically probe the interior of the living body. In one sense, then, one can say that therapeutic medicine, in practice, did not forget or abandon the lifeworld plenary-bodily mode of engaged knowledge. To the contrary, the highly honed skills of the surgeon who must feel as much as see the making of an incision remains very close—although in a specific and acquired set of bodily skills—to precisely the lifeworld of primary perception.

But does the science that gets favored, physics-astronomy, forget the lifeworld? My second foray again begins historically and reinter-prets another of the early modern science praxes precisely at its most visual moments.

I have already noted Galileo's fascination with his instrumentar-ium, primarily his telescope. What he was able to see with his tele-scope is what stood foremost in his interest. He couldn't wait to pub-licize his artificial revelation in his self-published *Heavenly Messenger,* in which he announced his various discoveries. But, like most science publications today, what gets publicized are the ultimate results, not the processes by which these are attained. In today's revolutionary or new philosophies of science—for example, with Kuhn and much more markedly with Latour—it is the process, including its failures, its movement from ambiguity to clarity, and its experimental develop-ment, which takes center stage. Phenomenologically, one can do the same with Galileo.

Derek de Solla Price, who sees much of the development of science in terms of its craft, the skills needed by the instrument maker, notes that Galileo's primitive compound telescope had such a narrow field of vision that to spot a mountain on the Moon was akin to seeing a faraway object by looking through two keyholes lined up a yard apart. What Galileo was interested in, of course, was the celestial phenomenon out there. His epistemology was externally oriented. And the optical mediation of the telescope dramatically modified what could be seen.

If we were to focus on the observational situation in terms of sheer visual and spatial aspects, we might note that the magnification of the Moon so that its mountains could be seen transformed much in relation to the eyeball observations heretofore possible. We could say the Moon became larger, magnified. But it was also displaced—telescopically it was taken out of the night sky and relocated within the field of telescopic vision. It lost its place in the expanse of the heavens and became a more focal, particularized object, now apparently close up.

But here already we can no longer remain merely visual in our analysis, because the apparent distance transformation via the telescope implies a change in apparent bodily position. We are, as it were, closer to the Moon, and it makes no difference whether we describe this as the Moon closer to us or us closer to the Moon. Indeed, the quasi-space of telescopic vision is itself a strangely transformed space. And it entails the phenomenological mutual correlation of thing seen with mode of seeing, now in the instrumental context.

The thing seen is, simultaneously, the same as anything seen without the telescope in that it occupies the same location in central vision and the same size of optimum visual distance (through focusing the instrument), and yet it is radically different from eyeball vision without the telescope. This phenomenon is today a virtual constant of more than visual experience. The technological near distance of the communications technologies (telephone, e-mail, televisual communication, conference calls) is a familiar new near-distant space.

What the story often neglects to tell us is that Galileo noted that he had to both tune his eyes to the instrument and instruct the unlearned how to use it (albeit he often put this in terms of treating his body as a quasi-machine itself). For the phenomenon of magnification is not monodirectional, but reflexive. Our two-keyholes-a-yard-apart description, however, shows this. The crude telescope, still lacking a motion-fixing machine that would eliminate or dampen the earth's and the body's motion and thus "fix" the object to be seen in stasis, magnified Galileo's own minute bodily movement just as much as it did the Moon object. Galileo had to learn to compensate for this by using a tripod, and by careful, and sometimes consciously developed, bodily motion. What the telescope magnified was thus that which was out there and that which was here, and the object seen and the way of seeing through one's kinesthetic body yield both a sense of the technological transformation of vision and of the reflexive correlation of seen-seeing. Moreover, one cannot speak here of anything like a monosensory phenomenon, but one must speak of a plenary one. Present at the first sighting of the Moon was the full, but now transformed, lifeworld of body-world correlations. What mediated and constituted the transformation was the instrument, the embodied technology of science. But that part of the actual praxis of science seldom enters the story and is forgotten as well by Husserl.

However self-interpreted, science in this praxis was fully multisensory and embodied in its observation, even though its observation was not direct, but mediated instrumentally. I have called this essential activity the instrumental realism of science. It is this instrumental mediation that links the body and thus the lifeworld to what is perceived by science.

Early modern science, of course, did not learn all these lessons nor follow what could be described as the double trajectory opened up by the magnification phenomenon. Instead, it remained largely extroverted and interested almost solely in the out-there. What if other directions had been taken?

Detour: Late Modern Science

What I have been describing, particularly with the historical asides to early modern science, omits a series of very drastic shifts from within science itself. Again returning to the favored sciences close to physics and cosmological astronomy, what has preceded has been without reference to the rise of relativistic and quantum physics. This I shall call late modern science.

What is of epistemological interest in late modern science is the shift from a kind of naive objectivism to an almost quasi-phenomenological relativity. By this I mean that the act of observation gets reinterpreted. And it gets reinterpreted in such a way that at least part of the body-world correlation gets taken account of—the observer's action must be considered.

Einstein's famous moving train example is one illustration of this reflexive switch. If one is in a car of one train, looking out the window at another, the illusion of motion that can occur sometimes confuses one over which train is in motion: is it mine or the neighboring train? Einstein's point, of course, is that all observed motion is relative to the position of the observer, and all that can be measured is the relative motion between the observer's position and that motion out there. In principle, while this is a quasi-phenomenological advance over any absolute and thus naive physics of space and time (such as Newton's), it could still be monosensorily visualist and thus still partially forgetful of the plenary embodiment of the observer.

In the actual experience, the illusion that can lead me to think my train is moving rather than the one out the window is almost instantly corrected by the more complete kinesthetic experience correcting the illusion. Were the experience solely visual, this correction might not occur.

Similarly, in relativistic and quantum considerations, the embodied action and instrument must be taken into account. To place a thermometer in a liquid does not simply record the absolute temperature of the liquid—it changes it by whatever the difference is between the two objects now conjoined, even if the effect is on the magnitude of a butterfly effect. Both relativistic and quantum physics are thus reflexively correlational in an approximation to phenomenology. A wider retro-extrapolation of this perspective could thus find a relativistic Galileo learning as much about embodiment as about the celestial world that interested him.

Imaging Technologies and Virtual Reality

We are now finally in a position to approach the most contemporary development of new instrumentation, in particular the proliferation of imaging technologies and of virtual reality developments as they bear upon science.

The imaging technologies are the most highly developed. I have already referred to some of the very sophisticated imaging technologies used in space instrumentation. The Venus probe radar scans are, of course, supplemented by a whole spectrum of other satellite and deep space instruments, most of which yield, again, visual displays. But some of these are particularly phenomenologically interesting.

Most imaging technologies are designed to retain what could be called perceptual isomorphism, that is, the display shows spatial and topographical features that, although often in black and white, still look like what one would see were one in the apparent position of the picture taker. Thus one can easily spot in the usual visual gestalt way the patterns that are craters or mountains or volcanoes. Perceptual reasoning includes this pattern recognition within the skill and experience levels appropriate to such identifications.

But isomorphism can be varied, minimally in one sense by image enhancement (usually through computers), which highlights contrasts, exaggerates certain features, etc. This might be considered a phenomenological variation, but note it is very like literary development as well.

The introduction of color—usually false color—further varies the perceptual gestalt of imaging technologies. In some cases the features that emerge are not available to eyeball perception at all, not by virtue of spatial distance but by position on the color spectrum. Infrared photography that highlights organic matter lets the observer "see" where vegetation is even from satellite distance. Stretching the variation farther, heat or light enhancement techniques provide yet other usually invisible features—the exhaust shadow of a recently exited jet on a runway is one example, and the ability to see in the night is another.

Even further away from isomorphism is the use of such deliberately variant optical imaging as spectrographs to determine the chemical composition of stars or other celestial entities in deep space. Here the rainbow configuration of lines is "read" by the observer, and any isomorphic shape disappears entirely.

One could expand upon the variants of imaging technologies in late modern science, yet as radical as these instruments are, they obviously continue both the preferred visualism of early modern science and the display of two- or three-dimensional patterns of early modern representation. In medical imaging in which finally an invasion into a living body can be made without serious danger or damage, the use of CAT, MRI, sonographic, and other imaging yields a look into the body itself, often in real time but usually in the thin depth of most optical instrumental displays.

Supplementing sophisticated imaging technologies today are the virtual reality developments in simulation. These developments are particularly interesting because the as-yet-undelivered claim is that the experience will be a plenary one, a whole-body experience parallel to real life. To be sure, this development is not yet fully formed, but could virtual reality instrumentation transform scientific instrumentation?

Let us note a few examples and their phenomenological implications. Perhaps the most sophisticated imaging technologies today are virtual reality simulators used by the military and aircraft industry, particularly for flight training and testing. My daughter-in-law flies Boeing 747s and annually has to undergo simulator training with lifelike situations of extremity one would not like to actually face. Pilots, even knowing it is a simulation, come forth sweating. Here the "realism" is multisensory and a higher quality virtual reality than most games.

In more recent times, the sheer speeds and complexity of supersonic fighter planes made very explicit the embodiment of pilots who were not merely visualizers but displayed very finite and limited powers as controllers of these high technology airplanes.

Epilogue

The itinerary I have taken not only follows what comes from embodied, perceptualist phenomenology, but identifies it within the context of an instrumental realism that links the lifeworld of embodiment with the farthest reaches of the micro- and macrophenomena that interest science. This trajectory is not the only one in which science gets at its phenomena, for I have not examined a much more hermeneutic route that often parallels the perceptualist one I have taken.

"Reading" instruments that yield nonisomorphic results, for example, data in the form of numbers, is obviously more hermeneutic in form. Its referentiality is more textlike than the direction I have taken. And yet both trajectories are ultimately complementary, as variations upon the things themselves. I have returned to one of the most basic origins of scientific knowledge that, through instrumental embodiment, brings back the lifeworld right at the center of frontier research.

Bodies in Science Studies

Part III

Chapter 5
You Can't Have It Both Ways
Situated or Symmetrical

Today's world is one in which we are frequently reinventing ourselves. I first invented myself as a philosopher of technology, then, finding that the nexus between philosophy of technology and philosophy of science might well revolve around science's technologies, instruments in particular, I began to move into philosophy of science. There I met, in reading and later in person, Bruno Latour and Donna Haraway and, a little later, Andrew Pickering. Popularly, they are frequently identified with what is today called *science studies*.

What was appealing was the way each seemed to have a sensitivity for the concrete, the material, which I found usually lacking in philosophy of science. But on closer reading, it soon became apparent that there were some issues that seemed to me to be incompatible—I phrase these for this chapter as a tension between *situatedness* and *symmetry*. I will look at that tension in science studies, in terms of the types and styles of ways in which the material world is dealt with, through various versions of hybrids, in this chapter focusing on cyborgs (Haraway) and machinic agency (Pickering) and my analysis of human-technology relations. A major issue that I see deals with the ways in which the analyses can proceed. On the one hand, this postmodern era is one in which the emergence of situated knowledge has become prominent and self-conscious; on the other hand, there has been an affirmation of various symmetries that purport to equalize the accounts of the nonhuman and/or material agencies in culture and especially technoscience. I shall look at both of these directions and argue that one cannot simultaneously be situated and symmetrical. In this context I will develop a thesis about what can be had, that is, I want to turn to some considerations of how one can responsibly deal with the material dimensions of culture and technoscience.

Postmodern Knowledges

To use the plural for knowledge, "knowledges," initially sounds a bit strange to anglophone ears. But it is more accurate today to describe what once was "Knowledge" as "knowledges" since one of the features of postmodernity has been the deconstruction of transcendentals and foundations, and replacement by local knowledges and particularized knowledge practices. Admittedly, this is not without contestation as the "science wars" arguments amply illustrate, particularly in North American contexts. Even from the bowels of my computer there is contestation, since whenever I type out "knowledges" the red-underlined misspelling warning appears in my text!

What has led to this change from Knowledge to knowledges? One factor relates to a major theme of my title, "situated" knowledges. Knowledges in the lowercase are *situated*. But what is it to be situated? My answer will be in the form of a narrative from my preferred philosophical framing practice of existential and hermeneutic phenomenology. In this tradition, to be situated entails that the knower is always *embodied*, located, *is a body*, and this must be accounted for in any analysis of knowledge. On one level this might seem almost too obvious—except for the fact that so much of the history of epistemology is one of different attempts to *disembody* the knower or to hide his or her embodiment. In my narrative I shall attempt to show not only the invariant role of embodiment in situated knowledges, but how "the body" is merely hidden in those epistemologies that attempt disembodiment. Then, in a second sense, one is also situated by cultural particularities which "mark" one's embodiment. These, too, must be taken into account.

On a second level, the knower is presumably working within some notion of the postmodern or amodern context that presumes overthrowing, surpassing, or moving beyond the modern. And historically, epistemologically, this is certainly the case—there has been in this century a variety of attempted overthrows of Cartesianism, which is often taken as emblematic of the modern. Indeed, both versions of the situated and later of symmetries are variations upon this rebellion. With respect to anti-Cartesianism, I shall follow the phenomenological revolt that deconstructs what Haraway calls "god-tricks" by recognizing the role of embodiment.

The third level or factor is the new, unique one, the role of the material: machinic agency, cyborgs, or human-technology relations, all attempt to incorporate material and nonhuman entities and the animal into the situation. There is a long story here that is about how the material world has also been left out of accounts of knowledge. The group discussed here, Donna Haraway, Andrew Pickering, and I, each try to reincorporate this dimension of existence, and my account must deal with this level as well.

So now, with respect to situated knowledges, the task will be to combine all these dimensions, levels, and factors into some kind of exemplar that can produce a gestalt. The devices I shall use are those that bring human knowers into intimate relations with technologies or machinic agencies through which some defined model of what is taken as knowledge is produced—I shall describe *epistemology engines*. My devices will be particular machines or technologies, which provide the paradigmatic metaphors for knowledges themselves. And through these narratives, I shall trace the visible and invisible roles of bodies.

Bodies

Donna Haraway has identified herself as one "who breathes Darwin and Foucault in with each breath." I shall duplicate this breathing by doing it with Merleau-Ponty and Foucault. Earlier I referred to body one and body two, twinned senses of bodies that owe their core significations to these two authors respectively. Body one is the existential body of living, here-located bodily experience, the sense of body elicited by Husserl as *Leib*, but much better descriptively developed by Merleau-Ponty as the *corps vécu*. Body one is the perceiving, active, oriented being-a-body from which we experience the world around us. It is the experience-as-body that is a constant of all our experiencings. (Body one is *not* the object-body subsumed under the mechanical metaphors of Cartesian early modernity—more of that later.) But, and I note this only in passing for now, this meaning of body is not directly or introspectively grasped—rather, it is interactively grasped by way of and in relation to the experienced environment or environing world. Its sense must be *reflexively* recovered. Phenomenologically one does not immediately apprehend that one's vision is perspectival, rather, one's invariant perspective on the world is reflexively realized by noting the ways in which that world "points back" to the null point of one's bodily position. In this sense, I learn my embodiment by actively being in a world. Body one is the necessary condition of all situated knowledges—but it is not the sufficient condition.

Body two is what could be called, out of context, the cultural or socially constructed body. It is the body of the condemned in Foucault, the body upon which is written or signified the various possible meanings of politics, culture, the socius. And it is the body that can have markers. It is the body that can be female, of a certain age, from a certain culture, of a certain class, and thus have a *cultural perspective* as the embodied and enculturated particular being we are. One can recognize here an aspect of Donna Haraway's version of situated knowledges. Yet I will reaffirm that for there to be a marked cultural body, or body two, there must be a body one that is markable.

What could be missed is the doubled deconstruction of modern epistemology that arose from both of these body discourses. A first deconstruction arose from the anti-Cartesian, embodied versions of knowledge developed in phenomenology. Husserl's presumed forgetfulness of the lifeworld by modern epistemology is a forgetfulness that ignores or overrides plenary perception, the sensory opening to the world from which all subsequent constructions are the second orders; the same goes for Merleau-Ponty's primacy of perception that is the originary opening to the world. In both cases the secret of body one is the clue to the deconstruction of Knowledge. For humans there can be no god perspective, only variations upon embodied perspectives.

The second deconstruction, although not limited to postmodern feminism, perhaps has its most virulent form there. One direction in this genre is to follow an almost total body two direction, but I shall not go that way. Rather, there is another strand that recaptures the phenomenological sense of body one while also accounting for body two results, for example, the group of American feminists such as Iris Young, Susan Bordo, Carol Bigwood, and others, who explicitly have drawn from Merleau-Ponty and the phenomenological sense of embodiment. Young's trilogy, "Throwing Like a Girl," "On Pregnant Subjectivity," and "Breasted Being" is a model development that recognizes both dimensions of bodies one and two.[1] Bodily motion, pregnancy, and breasts are real in both bodily-physical senses and the sociocultural senses that situate these phenomena. In one of my favorite Haraway essays, "Situated Knowledges: The Science Question in Feminism and the Privilege of Partial Perspective," she does what I recognize as an excellent phenomenology of instrumental embodied vision. While deconstructing the god-tricks of the views from nowhere and everywhere, she shows how "the 'eyes' made available in modern technological sciences shatter any idea of passive vision; these prosthetic devices show us that all eyes, including our own organic ones, are active perceptual systems, building in translations and specific *ways* of seeing, that is, ways of life."[2] Here is the crossing of bodies one and two, in the nonneutral position of situated knowledges. Both body one and body two underline situatedness.

What the body discourses contribute to situated knowledges is both deconstructive and reconstructive. What is deconstructed is the disembodied, nonperspectival, god-trick epistemology of early modernity. What is reconstructed is the sense of located, perspectival, embodied, and enculturated knowledge that is a praxis and action within and in relation to the surrounding world.

Epistemology Engine 1: The Camera Obscura

I shall now turn to a dramatic example that links bodies, technologies, and their interactions such that an epistemology is invented in Foucault's sense of an episteme. This set of practices revolves around the paradigmatic role of the camera obscura in early modernity. "Camera obscura" literally means "dark room"; it is a device that captures an optical effect:

> This "dark room" has a small opening on one side and a blank (preferably white) wall on the other. When the lighting outside is proper, an inverted image of objects or scenes outside the camera are cast upon the blank wall. This inverted image phenomenon may have been known to Euclid, but it was an effect deliberately employed and described by Alhazen, the Islamic thinker, who used it to observe eclipses and who described it—along with an analogy between the *camera* and the eye, in 1038.[3]

It was rediscovered in the Renaissance and became one of the favorite technological toys of this vision-obsessed period. Leon Battista Alberti (ca. 1450) used a "camera" to produce "wonderfully painted pictures of great verisimilitude."[4] Here was one of the first draw-by-the-lines techniques and one of the techno-elements involved in the Renaissance rediscovery of three-dimensional perspective drawing. While other devices were also used, the camera was one of the best to "automatically" reduce three-dimensional objects to two-dimensional images.

Leonardo da Vinci (ca. 1450) was also fascinated by this toy and he reinvented the camera/eye analogue:

> When the images of illuminated bodies pass through a small round hole into a very dark room, if you receive them on a piece of white paper placed vertically in the room at some distance from the aperture, you will see on the paper all those bodies in their natural shapes and colors, but they will appear upside down and smaller.... the same happens inside the pupil of the eye.[5]

This optical toy continued to be of use into early modernity. Galileo used a variation of it to discover sunspots (1630)—a telescope plus a helioscope. But what could be missed is that the camera obscura played a very explicit role in the modeling of modern epistemology itself. It was explicitly used by both Descartes (for the Continental rationalists) and by Locke (for the British empiricists). In La Dioptrique, Descartes notes:

> If a room is quite shut up apart from a single hole, and a glass lens is put in front of the hole, and behind that, some distance away, a white cloth, then the light coming from external objects forms images on the cloth. Now it is said that this room represents the eye; the hole the pupil; the lens, the crystalline humor—or rather, all the refracting parts of the eye; and the cloth, the lining membrane, composed of the optic nerve-endings.[6]

But what Descartes adds is another analogue: the mind or mental substance—the modern *subject* who, as we will see, is now inside the *camera*. Lee Bailey points this out in a very important article, "Skull's Darkroom":

> The *camera obscura* began as an experimental model for the eye and became a ruling metaphor for the mind. By offering a way of picturing the Cartesian inside *cogito* with a sensory channel admitting pictures from the outside *extensio*, the image of skull's darkroom shifted from a suggestive experimental analogy to a concealed methodological paradigm.[7]

In short, the modern subject is the homoculus inside the camera obscura. What comes from the outside are the impressions from the *res extensa* that are cast inside the box or body upon its receptor, the eye (retina) analogue where *images* form that *represent* the external world. So here, at a stroke, we have invented early modern epistemology with its (a) individualized subject (b) enclosed in an object-body, (c) thus creating a body/mind dualism (d) with no direct knowledge of the external world, but a representational one by way of images.

Nor is this early modern model of knowledge limited to Descartes; it is even more explicit in Locke:

> *Dark Room*—I pretend not to teach, but to inquire, and therefore cannot but confess here again, that external and internal sensation are the only passages I can find of knowledge to the understanding. These alone, as far as I can discover, are the windows by which light is let into this dark room: for methinks the understanding is not much unlike a closet wholly shut out from light, with only some little opening left, to let in external visible resemblances, or ideas [images] of things without: would the pictures coming into a dark room but stay there, and lie so orderly as to be found upon occasion, it would very much resemble the understanding of a man, in reference to all objects of sight and the ideas of them.[8]

Once again we have the camera as the explicit model for knowledge with the equivalence now of Locke's famous tabula rasa as the white screen upon which are cast the representations of things in the external world. The camera model belongs to both sides of the English Channel in early modernity. The subject (self inside the camera) can only directly be aware of the images (representations) of the external world cast upon the white surface inside the camera. Herein lies the "invention" of the modern subject or cogito.

It is this subject that is said to have disappeared from postmodernity. (I shall show you, shortly, that it has not disappeared, particularly within the cognitive and neurosciences of the present that still operate in the modern, but more of that below.) If the modern subject has disappeared, what caused its demise? My story now returns to the phenomenological deconstruction performed upon the cogito.

What I left out of the camera-modeled epistemological frame was the problem it created for its inventors. If all we can know (directly) is the image or representation of the external *(res extensa)* world outside, how can we know that there is an accurate correspondence between image and thing? And we all know Descartes's answer: it is the epistemological god who guarantees this correspondence.

Here, then, is the early modern invention of the god-trick, the nonperspectival perspective of the ideal observer who rides above the world. God sees, but because of being god, his view is presumed nonperspectival. I will not bore you with the centuries of debate about this, but I will pull a simple existential phenomenological trick to deconstruct this picture of knowledge: *where is Descartes when he makes the claims about knowledge that he makes?* The answer is outside the camera, viewing it from both the outside and the inside. Descartes himself is the "secret" of the god-trick since in his nonspoken, privileged position he can "see" if there is correspondence or not. Descartes is using what today in computer games is called a cheat code. He already knows the answers from his privileged position or perspective.

Nor shall I here take you through all the complicated steps Husserl took to show precisely this, but jump to the conclusion: Descartes's "cheat code" position is not *transcendental*, but embodied. It is the "here" from which both camera and homoculus, and its external world, is seen. Phenomenology takes the subject out of the box and places him or her *in the world*—but as embodied and in a perspective. Merleau-Ponty makes the point more succinctly: "Truth does not 'inhabit the inner man,' or more accurately, there is no inner man, man is in the world, and only in the world does he know himself."[9] I am the situated "where" from which I see—experience—the world. "The world is there before any possible analysis of mine,"[10] but this primacy of the world is simultaneous with the positionality that I am as body: "I am conscious of my body *via* the world. . . . I am conscious of the world through the medium of my body. I am already outside myself, in the world."[11] Thus phenomenology takes the cogito out of the camera and finds it, embodied, in a world. This reframes the entire context of the camera. But one must also note that the price for this liberation is one that now entails the perspectivalism of embodiment rather than the nonperspectivalism of the god-trick. Surely this is a situated knowledge.

The tactic that I have just displayed utilizes an artifact (a machinic agency) that was taken as a paradigm for the human activity of making knowledge (a cyborg activity), to produce a picture and thus self-understanding of that activity. And, albeit in too simple a way, I used the invariant of embodiment to both construct and deconstruct this invention of modernity. I underline only two points in this example: (1) By taking the subject out of the camera and finding him or her *in the world*, one simultaneously has deconstructed the god-trick of early modernity and established an embodied situated knowledge. This post-Cartesian embodied being, with perceptual perspectivity, is an invariant of such situated knowledge. (2) Then, retrospectively, we can also see that the cogito or homoculus-subject in the camera was a "machinic fantasy," an artifact of the device used to model knowledge.

We have now rearrived at the postmodern or amodern that has proclaimed the cogito dead or never existent. But this "post" position remains contested because in the contemporary world, and most particularly in the neuro- and cognitive sciences, the camera model still holds sway and has even renewed itself.

What we have here is simply the latest version of the camera. The fMRI imaging has moved the location of the mind's image from the pupil (da Vinci) and the retina (Descartes) to deeper inside the body-box to neural brain activity. The subject still reports as the privileged homoculus inside the camera. In short, the science wars are not yet over.

Semiotics and Symmetries

I now turn to the second postmodern revolt against modern epistemology, semiotics, and the symmetries that result; within science studies this issue has taken on something of a life of its own. Pickering's book *Science as Practice and Culture*[12] contained the infamous chicken debate. This debate, which still reverberates, was a battle between the UK-based sociologists of science from the strong program and Bath school of interpreters of science, and the Continental, mostly French, "actor network" thinkers arising from the work of Michel Callon and Bruno Latour. The UK sociologists held that there is a priority to the subject, a sort of anthropomorphic centeredness, that could not be overcome. The actor network people held to a stronger symmetry based upon semiotic principles that, in effect, saw both humans and nonhumans as "actants." The chicken debate was about the escalation of symmetries that are found in various forms of the sociology of scientific knowledge (SSK—anglophone) and science studies or sometimes the study of science and technology (STS—francophone) and in between. The weak symmetries of the strong program and Bath school were largely framing symmetries that rejected distinguishing "true" science from "false" science in advance and instead concentrated upon how one socially produces agreed-upon results. Its weakness, according to Pickering, is the absence of machinic agency. But this debate itself recognizes stronger symmetries that arise out of Continental *semiotic and structuralist* traditions, associated more with actor network theory approaches—but also now claimed to some degree by Haraway and Pickering. I shall focus upon this stronger version of symmetries and relate them to the move into the postmodern or amodern.

Semiotics originates in a general theory of signs, and as Collins points out, ultimately it "treats the whole world as a system of signs . . . [and] where there are no differences except differences between words there are no surprises left."[13] But how does this relate to symmetry? Semiotics, historically and philosophically, belongs to the family relations of structuralism, general linguistics, and semiotics, as noted. This is the family of theory framing that immediately preceded poststructuralism, postmodernism, and deconstruction. Its ghosts and its poltergeists still haunt the current rage of francophone-influenced contemporary thought.

To understand what is involved, I shall look at semiotic ancestry. The crucial point of origin was Ferdinand de Saussure and his theory of general linguistics. At its core was a distinction between *langue* [language] and *parole* [speech]. *Langue,* or the system of language, was conceived to be a system of signs, finite in number, but infinitely combinable to make up the range of linguistic meanings. Included in this concept are two modeling ideas. In the first, language becomes a system, closed and finite with respect to units, an object (or object of study). This is a framing device, and all theories do something of this sort—one never studies a naked object, but frames it and places it into a manageable context. This is what Galileo did to motion by framing it onto inclined planes or dropping objects off the Tower of Pisa. But to frame is also to *transform,* or in today's language, to *construct.*

The second modeling idea is one in which *writing,* even better, *alphabetic* or *phonetic* writing, is secretly imported into *langue.* For linguists, words are constructed of phonemes, which, while not identical with alphabetic letters, are still sound bits and finite in number, thus subsumable under the notion of system above. This move accomplishes two things simultaneously—it "deconstructs" speech into units, and it "subsumes" living speech into a position within language-as-object. Speech becomes, simply, the set of possible moves within language, and thus language becomes the "transcendental" dimension of all possible signification. This doubled systems notion of language grandfathers semiotics, structuralism, and poststructuralist deconstruction.

I shall not trace out developments in structuralism and later in poststructuralism, but turn immediately to the moves that help establish the framing tactics for *symmetries* of the strong variety.

First, structural linguistics-cum-semiotics makes possible a different version of the deconstruction of the subject. In the weak sense, speaking subjects are seen from this frame as simply the instantiators or operators of some set of linguistic or semiotic possibilities. Whatever can be said, can be said meaningfully only within the system of language. And while this move deconstructs—as did phenomenology earlier—the Cartesian spectator consciousness, it now drives whatever vestige of subjectivity there could be in the direction of linguistic-like signifying activity. Thus, from my emphasis upon an active, perceiving embodiment, this is now a non-Cartesian form of disembodiment.

Second, a structural semiotics makes possible the leveling of dichotomies and distinctions, or a radical symmetry. All significations are ultimately merely the transformations of signifier-signified. In the contemporary debates there are some interesting kinship lines here. For example, Derrida takes his poststructuralist direction into the realm of texts and inscriptions—the moves of radical symmetry horizontalizes texts, with margins, glosses, footnotes, and even the blankness of the pages symmetrized. Or, with our absent Latour, one can have actants, whether human or nonhuman, move equivalently within the structural system. One can study these transformations, of course, but in a sense, they simply occur. As with the non-Cartesian equivalent of disembodiment, this radical symmetry reintroduces a new, non-Cartesian equivalent of the "view from nowhere." Who describes the symmetries?

Third, semiotics—one can say positively—"textualizes" the world. But, negatively, this is also to *reduce* the world to a language-like being. There is something of an inversion here from the "world of the text" into the "text as world." From this perspective, if one wants to maintain an emphasis upon practice, action, and production, then one sees the actions being performed as those that produce *inscriptions*. We are again in a non-Cartesian mode, but a mode in which there reemerges another version of the reduced world of the *res extensa*. The semiotic version of the *res extensa* is this world-as-text read by an invisible reader.

I wish to make one final observation about semiotic methods. While these are usually associated with the postmodern, there is also something quite premodern about them. As Foucault and others have noted, one of the changes between the premodern and modern epistemes (epistemes are discrete periods of knowledge formation, according to Foucault, and are chronological versions of the knowledges that have been discussed above) consisted of the change from all meaning being essentially found in language to meaning found through perception—Foucault goes so far as to claim perception is invented by modernity. But with the linguistic turn of semiotics, we return to this dimension of premodernity. Of course, there is a difference: semiotics is a code more than a language, and its applicability to everything from computers to genetics draws upon this trope.

I now have situated the contraries between situated knowledges and symmetries, and, as my title claims, I must show why "you can't have it both ways."

Slippery Symmetries

Semiotic-based symmetries have historically taken a number of directions. The early Derrida moved to invert what he took to be the phenomenological primacy of speech over writing or inscription; this move could be seen as a simple variant upon structuralist linguistics. The once faddish structural anthropology of Lévi-Strauss in particular—far more than the "modernity which never was" for Latour—escalated the nature/culture distinctions (the raw and the cooked, etc.) that were semiotically interpreted in terms of a binary set of codes enclosed within "primitive" myths. But in our context, the symmetries that are relevant are those that place into a neosemiotic system humans, machines or technologies, and animals as well. While I shall continue to affirm—and join—the attempts to incorporate and take into account the nonhuman agencies with which we interact and which are integral to understanding the contemporary world. I begin with a short critique of symmetry strategies and what I take to be their weaknesses.

If one places on a continuum a number of varieties of symmetry strategies, what one finds is that old-fashioned, modernist reductionism is a type of symmetry. It employs uniform rhetorical and linguistic descriptions. Its voice is the well-known anonymous voice of technical writing that (a) employs rigorous avoidance of all anthropomorphisms, (b) is usually cast in terms of formal or abstract or, even better, mathematical formulations, and (c) reduces all entities to variables within its system. Its usual form is that of physicalism, illustrated above in the brain states fMRI example I gave.

At the extreme opposite of this continuum, one finds what I shall call a new social-anthropomorphic rhetoric and reduction. Michel Callon's infamous study of the scallops of St. Brieuc Bay are illustrative. Here we indeed find the scallops as actants or agents in the project, but as agents, they are described in social-anthropomorphic terms. The scallops take on quasi-intentionality and actions. Once again, we have a uniform ontology, which while now anthropomorphized is nevertheless simply an inversion of physicalism. Note that with both physicalist and anthromorphist strategies, one has a uniform ontology that easily lets all the actants operate within this now united system of variables. What could be called uniform ontology symmetries can easily have substitutable variables, but at the price of a reduction in ontological variety.

In between are the symmetries that might be called composite or hybrid symmetries—and here is where I place Pickering and Haraway. These hybrid symmetries want to have it both ways. Although much more complex than I shall describe it here, Haraway's strategy is to string whole groups of heterogeneous elements into a single, complex unity—"seed, chip, gene, data-base, bomb, fetus, race, brain and ecosystem" united under the coined hybrid term "cyborg" or cyborg figure. This rhetorical device presumably allows the complexity of contemporary relations to be shown while not reducing the elements to a singular ontology. The other element of Haraway's neosemiotics is not reductionism, but boundary blurring. OncoMouse is simultaneously nature/culture, constructed/born, human/animal. This "both/and" rather than "either/or" is a mark of postmodernity. What should be noted here is that the power of this device draws simultaneously from ordinary meanings and from systemic transfers, i.e., having it both ways.

Pickering's approach is one that is subsumed under his metaphor of the mangle, which, like the old machines that ironed the laundry, combines together humans and machines in practices or performances. There is a richness of performance analysis in the mangle, with human agency coming up against resistances and accommodations in a dance of agency. But while extolling symmetry, Pickering hedges his bets by retaining intentionality only for human agents, although allowing that machinic agency and human agency are in some ways interchangeable in the mangle—again, having it both ways.[14]

I argue, of course, that such symmetries revert to functional equivalents of precisely the Cartesian modernism that postmodernity wishes overthrown in that (a) the perspective from which the symmetry is drawn is unknown, (b) the absence or transcendence of the narrator again creates a god-trick of nonsituatedness, and (c) the question of for whom the system operates also hides the politics of semiotic systems. With my neosemiotic colleagues, there are weak attempts to address this. Haraway, in her first-person narrative, inserts herself as situated within the cyborg context—she admits her motives and marks and speaks these out as part of her political program.[15] Pickering, still in the quasi-anonymous third person, keeps situatedness within the mangle by retaining intentionality as planned and motivated actions with the humans. Thus, I conclude that neither Haraway nor Pickering are fully or genuinely symmetrists but are, at most, quasi-symmetrists.

So, what do we have?

Epistemology Engine 2: Cyborg-Cyberspace Technologies

Imagine now a new machinic context, this one more complicated—in keeping with our postmodern situation—than the camera of early modernity. I am going to use here a series of related technologies that incorporate computers, networking, and multimedia taken to virtual reality developments to invent a new episteme. From these *epistemology engines,* as with the camera, I shall analyze a postmodern or amodern subject who may or may not be cyborgean.

Variation 1: Video and Computer Games

My first example comes from video and computer games. Here our human is seated before a screen and *interacts* with it in terms of a joystick. There is even a superficial similarity to our previous camera in that the action displayed is on the screen (or, for cybernauts, through the screen). But there is also something radically different here— there is no question of a correspondence between the screen imaging and the real world. The world-on-the-screen is a fictive world that is constructed, not copied. And while it may mimic the medieval, the animal, more likely the science fictional, it is the imaginative invention of the absent programmer. In my earlier human-technology relations terminology, this is an *alterity relation* in which the machinic entity becomes a quasi-other or quasi-world with which the human actor relates.

My adolescent son (thirteen years old) gets obsessed with each new computer game: "Awesome graphics, Dad!" And for some period of time, until all the levels are mastered, all the lives lived, and the ultimate enemy faced—"Diablo, at last"—the game remains a fantasy enchantment. Here is one secret to this new device—it has a *fantasy trajectory* by its *virtuality.* I watch him play at his request, and with boredom I wonder how he can be so enchanted when the varieties of chase-and-kill are simply those of being a warrior with a sword or, in a new graphic, a soldier with a machine-pistol, or a science-fiction hero with a phaser. The genres are limited as well, to puzzle games where secrets must be revealed (usually with cheat code books), hunt-and-chase, or construction "sim" games.

Of course, as soon as he walks away, the virtuality is turned off, and while there may be a carryover (the cheat code and manual follow him to the breakfast table), he is more fully motile than the previous eye-hand coordination action, and there is not here the contemporary slide from RL to VR. But there is one preliminary background note about the perspective that is worth note: many of the games come with alternative perspectives built in. In Flight Simulator one can fly the virtual airplane from the perspective of the pilot. Or one can switch to the third-person perspective and see oneself as a quasi-object in the airplane now over there approaching the runway, the building, or whatever. There is an easy switching of perspectives, made equivalent and thus only quasi-embodied in the game.

Variation 2: Computers and the Internet

I now switch to another interactive device, the computer networked through e-mail. We now have at least two persons wired in a connection. This time, the focal mediation is textlike and the screen shows print messages. In my example, I take a mediated communication between two persons who have not actually met face-to-face. This is a *virtual* meeting and communication. It falls under what I have earlier called a *hermeneutic human-technology relation* in that the machinic mediation presumably refers to a real other of some sort and becomes in this case a kind of language-analog mediation. But again, we are no longer in the Cartesian camera situation because while one can raise the question of the correspondent reality of the other, this mediation is one that remains virtual in context.

In ordinary social contexts, the bulk of such connections probably remain mundane and do not tempt one to raise critical questions about virtuality. But increasingly there are e-contexts in which the *virtuality trajectory* does take place. *Science* describes a new study on the Turing Game, a game in which the interlocutors interact and try to infer the real identities of each other, but only at the end of the game do the interlocutors "confess" these identities. How do I know the other is male? female? young? old? heterosexual? gay? white? of color?[16] Virtuality here plays a masking role or, at the very least, a self-selecting role. This fantasy game, showing the fantasy trajectory noted above, can go in other directions as well—for example, journalists have become fascinated with the growing phenomenon of the e-romance. In this variation, our interlocutors become virtually romantically enchanted. The ultimate results of such virtual romances have sometimes been those of disenchantment upon real meetings or even real divorces based upon virtual adulteries. Again, the new devices enhance the possibilities of fantasy and virtuality. These are not primarily correspondence epistemologies.

Variation 3: Multimedia and Virtual Reality Technologies

The previous machines are obviously reductively limited—in the first case the eye-hand coordination, supplemented by audio sounds, still restricts full bodily movement and perceptual richness. In the second case, the silent texts of e-correspondence, while allowing all the imaginative expansion and variation that any literary vehicle can develop, remains even less embodied, although it, too, can be enhanced. The contemporary technologies in this variation are degrees of such enhancements.

Let us return to our e-lovers. As they become more enchanted with each other, the fantasy desires that e-communication enhances can be adumbrated with, for example, voice messages and digital photographs. And while, maximally, these could be just as selective or even false, they are likely to be at least minimally self-flattering. Our e-lovers do not send each other nonflattering digital photos. (Here I take a small detour to show the same effect. In a paper by Philip Brey there was a description of the Dutch Philips Corporation's attempt to invent technologies that would preserve or enhance quality-of-life devices.[17] For example, one idea was to provide children with "emotion sending beepers," beepers that, if the parent knew an exam was to be taken, could send a message of encouragement by making the beeper become warm—warm beeper equals warm feelings. But one puzzle that Philips Corporation encountered was why the audio-visual telephone had not caught on. Their market researchers found that many potential customers disliked the idea of a visual phone catching them unawares—for example, on the toilet, or not made up, or unshaved. Philips proposed to counter this by building into the technology a selection of avatars, that is, self-selected flattering images. When you ring and I pick up, the audiovisual phone projects me sitting comfortably in my study, all made up with proper hair and a nice turtleneck. So here we are, back to virtuality.)

What is the epistemology of these styles of virtuality? It clearly is not Cartesian since the design is one that enhances masking, fantasy, and a certain kind of construction. One could also call it a highly active and, to play upon one of Pickering's terms playfully, a perform-ance epistemology. But this kind of performance is more theatrical than realistic and with interesting implications for the subject. Our subject is obviously not the Cartesian cogito at all, although Descartes's concern about whether or not we could be fooled by a very cleverly contrived automaton does reemerge here in another form.

Rather, our new subject is one who in the *dance of new agency* can and does become a multiple-roled *actor*. In an earlier work I sug-gested that yet another variant upon these technologies, the multi-screen display, is suggestive. We enter into, in practice, so many vari-ations upon virtuality that we can, reflexively, begin to *edit ourselves in terms of multiple roles*. We can choose, as it were, from the video-game perspectives—were it not for our lived bodies—and enter any number of social and cultural roles that, bricolage-style, we can pick and choose, even in terms of culture bits, to edit our style of life, a multimedia style if you will. It is an editing or fashion style of existence.[18]

But we have one more step—the one that is so popular, the slip-page from RL to VR. This existential fantasy is—and I have a hard time believing this—sometimes taken literally. At its extreme, it con-sists of those who want to be "downloaded" into their computers or have their bodies "hardwired" into their computers—but if I am right about embodiment, the result would be a major shock. Short of be-coming our favorite machine, there is the lesser step of the styles of virtual reality machines that move from today's limited audiovisual technologies toward whole-body technologies that ideally would in-corporate the full sensory spectrum, especially tactile and motile virtual phenomena. (I have to say that those that I have experienced remain quite primitive and disappointing. The feedback from smashing ten-nis balls, the quasi-vertigo from free falls, indeed, the weirdly unreal sense one gets from these various enclosures, for me, enhances the strangeness rather than the presumed possible slide into permanent VR. Have you ever eaten a virtual McDonald's burger? Or better yet, imagine the fulfillment of VR sex—it is rightfully described as nei-ther hetero nor gay; the internal trajectory of VR fantasies may best be seen here since VR sex is necessarily masturbatory and narcissis-tic.) These machinic developments, of course, remain on precisely the trajectory I noted from our first example, a trajectory into the totally constructed, virtual world of the Total Machine. In this sense, we remain precisely on the larger trajectory of modernity in its later guise, toward another variation upon totality, now modified into a *virtual totality.*

Postmodernity as Another Machinic Fantasy

By now my very tone has given me away. I obviously believe that our current enchantment with this family of virtual trajectoried machines is just as much an enchantment with a machinic fantasy as early modernity found itself enchanted with its own epistemology machine, the camera obscura. Yet if we were to take my computer-cyberspace machines as epistemological engines, one could see what kind of subject they would have.

The virtual subject is multiple, not identical. As the avatar example indicates, there are many roles and personalities that the wired subject can take. Every new situation provides new relations and new possible identities. Cyborg identities are thus more like the multiscreen images in news rooms, and as individuals we can "edit" our beings by switching from one screen to another.

Positionality, which is a feature of embodiment, can in cyberspace be one of alternation in perspectives—this is already built into computer games, and the player can choose whether to be in the active-role position of the null-point player or take a more overhead position and be in a quasi-out-there location, taking the self-position as quasi-other.

As the fantasy element of cyberspace is amplified, one can choose to be anything one can imagine. So here the elements of fantasized cultural bodies come into play as a kind of instant machinic theatrical role. Boys become instant hunks and girls instant models, or, as movies such as *Lawnmower Man* fantasize, one can become anything imaginable in virtual space.

And the ultimate fantasy, of course, is to take the slippery slope of projecting RL into VR, the ultimate machinic fantasy. In the most literal sense, this fantasy is already desired with those who wish to be permanently wired, downloaded into their computers, becoming their machines—the ultimate techno-narcissism.

Epistemology Engines

While my playful epistemology engines have suggested forms of being, I remain skeptical and distant from this too-easy way to model knowledges upon the nonhuman machines we invent. I have suggested that I take both the subject of the early modern camera and of its postmodern equivalent in cyberspace machines as machinic fantasies. And while I am hopeful we can avoid the centuries-long captivity the modern subject had in its camera-box and find escape from the cyberbox we are now constructing for ourselves, there is something deeper lurking in these human-machinic encounters.

Descartes both saw himself and did not see himself in his camera, and I suggest we do the same with our cybertechnologies. It is my contention that machines, the animal, or border objects such as OncoMice and Asian eels do not show themselves nor do we show ourselves directly as representations or images or pure objects. Rather, it is in the interactions, in the mutual questioning and interacting of the world and ourselves, in the changing patterns of the lifeworld that things become clear. In early modernity, the camera—and the other visualizing technologies in the interaction of humans and non-humans—produced in action the then new way of seeing that became interpreted as modern epistemology. It was the action in the complex and changing *life*world that allowed this way of seeing to become stabilized. Similarly, in postmodernity, with a new set of "toys," cyberspace and computer toys, we are taking shape in a new set of relations such that both world and self take on different dimensions. But whatever these new realities are, they will emerge from the dance, the interrogation, the "foldings of the flesh" that Merleau-Ponty talked about in his late works, and they may be located by looking at the practices and in giving account of our bodily engagements and embodiments in that world. These are the directions that a critical phenomenological and ontological investigation of knowledges might take.

I am suggesting that while one can directly seem to take account of such features of the machinic as its technical properties, this is itself merely one interactive variant upon the material. Another, the line I have developed, shows the current machines to be fantasy-enhancing devices as at least another trajectory of their human-nonhuman relational being. In short, both human and nonhuman agencies get revealed indirectly, through the critical examination of the patterns of lifeworlds that indeed contain humans and nonhumans, even cyborgs. In this interconnection of embodied being and environing world, what happens in the interface is what is important. At least that is the way a phenomenological perspective takes shape. This theory of relations is one valid way of taking into account humans and nonhumans, but one that eschews ontological reductions, both naturalistic and semiotic.

Chapter 6
Failure of the Nonhumans
A Science Studies Tale

By the early nineties it was obvious that the demands of the nonhumans could no longer be ignored, and in this present decade, opening the millennium, their failure is equally obvious. This tale traces their attempts at being heard and recognized but, at the sad ending, that resulted in their continued silence and obscurity.

The tale opens with the shrouded invisibility of the nonhumans. What were to become characters in the later tale—cyborgs, hybrids, the scallops, silent policemen, NRA guns, instruments of science— were, if mentioned at all, regarded as neutral and passive or, better, totally transparent entities. They had not yet made their demands known and were, at best, nominal pawns to be played within the chess game of theory and human actions. Those who played the pawns were foremost the philosophers and sociologists of science and even of technology, but sometimes also the scientists and designers of the nonhumans themselves.

The nonhumans were, as this tale begins, most invisible among the tribes of humanists—that is, if the reports of their humanist defenders are to be believed. Bruno Latour, the primary etymological inventor of the "nonhumans," claims to have given them their due, particularly within the sciences: "Instead of the pale and bloodless objectivity of science, we have all shown, it seemed to me, that the many nonhumans mixed into our collective life through laboratory practice have a history, flexibility, culture, blood—in short, all the characteristics that were denied to them by the humanists on the other side of the campus."[1]

Nor is Latour alone in decrying the absence of the nonhumans. Andrew Pickering accuses his peers in the sociology of scientific knowledge of also ignoring them: "Talk of material agency has always been suspect in the sociology of scientific knowledge, but not so in the actor-network approach. There, much is made of material agency and, further, of its symmetrical relations with human agency."[2] Thus, in his accounts of laboratory life, Pickering needs to add what is missing:

"I could not persuade myself of the story that I wanted to tell of Morpurgo's early struggles to get his apparatus to work.... something seemed to be missing. And that something, it appeared, was material agency. In building his apparatus, Morpurgo was trying to get the material world to do something for him, and this needed to be stated out loud."[3]

Yet earlier and more radically than either, Donna Haraway proclaimed the nonhumans as part of the human/nonhuman hybrid under the sign of the cyborg. "Cyborgs [are] creatures simultaneously animal and machine, who populate worlds ambiguously natural and crafted.... the boundary between human and animal is thoroughly breached ... [and] the second leaky distinction is between animal-human (organism) and machine."[4] According to Haraway, we are all already cyborgs.

The tale now has its protagonists, the trio of Haraway, Latour, and Pickering, speaking out for the nonhumans against the traditions of the philosophers, sociologists, and humanists who have ignored them. Here we arrive at collectives that include nonhumans in symmetrical interactions with humans or with machinic agencies, including movements of "tuning," a "dance of agency" that includes resistance and accommodation, or with hybrid-cyborgs of various mixtures and ambiguities. Still, it should be noted that there remains no consensus about just who or what these nonhumans are. If we are truly hybrid-cyborgs, then we are them. But if we enter into a "dance of agency," they may not be us, and they may not even have intentionality, as Pickering claims: "I want to discuss an aspect in which the symmetry between human and material agency appears to break down. I want to talk about *intentionality* .... I find that I cannot make sense of the studies that follow without reference to the intentions of scientists ... though I do not find it necessary to have insight into the intentions of things."[5] Then, for Latour, the humans and nonhumans are or become interchangeable: "Realism gushed forth again when ... we began to speak of *nonhumans* that were socialized through the laboratory and with which scientists and engineers began to swap properties.... realism became even more abundant when nonhumans began to have a *history*, too, and were allowed the multiplicity of interpretations, the flexibility, the complexity that had been reserved, until then, for humans."[6]

In short, although the nonhumans have arrived, they have failed to establish a consensus among the humans about who or what they are. They are us? They are different than we are, but enter into "dances" with us? Or they are not us, but are fully interchangeable with us? Yet in all these cases, the nonhumans now appear to be actants of some kind. How did this happen?

If we read the histories by which the nonhumans began to show themselves, in contrast to their silent invisibility attributed to the philosophers, sociologists, and humanists, the event-origin seems to be the invention of the air pump in the seventeenth century as described by Steven Shapin and Simin Schaeffer in *Leviathan and the Air Pump*. This book becomes the point of origin that is widely commented upon by each of the trio of nonhuman defenders noted above.[7]

Whether or not there was a Scientific Revolution, there is wide agreement that there was a shift in what could count for knowledge during the seventeenth century. That shift carried what counted as evidence from the textual to the observational realms, from that which was written to that which was perceived. This states the matter too simply, but it does begin to indicate that what could be observed — and as we shall see primarily by means of technologies (a species of nonhumans) — began to count as what was real for early modern science. Galileo claimed that what the telescope showed exceeded the claims of Aristotle and the Church Fathers; the microscope's better revelation of the anatomy of bees by the observers of the Society of the Lynxes claimed the same thing. This same claim occurs in the machinic operation of the air pump. Each move and machine belongs of a piece to an epistemic shift. *Leviathan and the Air Pump* explicitly affirms that this shift is a result of fact production through nonhuman agency, the technology of the air pump.

Shapin and Schaeffer utilize a Foucault-like notion of interrelated technologies to describe the construction of facts within experimental life:

> We shall show that the establishment of matters of fact in Boyle's experimental program utilized three *technologies:* a *material technology* embedded in the construction and operation of the air-pump; a *literary technology* by means of which the phenomena produced by the pump were made known to those who were not direct witnesses; and a *social technology* that incorporated the conventions experimen-

tal philosophers should use in dealing with each other and considering knowledge-claims.[8]

But it is the nonhuman or material technology that is of primary interest here. The new material technology (air pump) was part of the early modern technological embodiment of scientific vision.

> The power of new scientific instruments, the microscope and telescope as well as the air-pump, resided in their capacity to enhance perception and to constitute new perceptual objects. . . . Hooke detailed the means by which scientific instruments *enlarged* the senses: . . . his design was rather to improve and increase the distinguishing faculties of the senses, not only in order to reduce these things, which are already sensible to our organs unassisted, to number weight, and measure, but also in order to enlarge the limits of their power, so as to be able to do the same things in regions of matter hitherto inaccessible, impenetrable, and imperceptible by the senses unassisted. . . . Scientific instruments therefore imposed both a correction and a discipline upon the senses.[9]

With the experimental life, matters of fact—perceived—could be produced through the actions of nonhumans or instruments. The very notion of objectivity takes on this sense of machine-produced facticity. This was the entry of the nonhumans into the domains of philosophers, sociologists, and humanists.

Yet this epistemic shift was not without contestation. For even if modest witnesses could testify to suffocated birds, rising and falling columns of mercury, and nonfluttering feathers inside the air pump, nothing was ever defined with certainty due to the ambiguities of pump leakage, failures of replication, and other machinic failures. Perhaps, at most, perceptual ambiguity replaced textual ambiguity? But the small gain of recognizing the explicit roles of the nonhumans now could not be ignored. This small gain, however, did not resolve the problem of a lack of consensus about who or what the nonhumans do by way of action.

If we now return to our trio of nonhuman defenders, we can discern different degrees of what can count as actions. The strongest stance is taken by Latour with his earlier notion of actor networks, later replaced by the notion of a collective (of humans and nonhumans). In both cases, we have positions of interchangeability of humans and nonhumans in a strictly symmetrical set of roles. The collective, for Latour, is the combination of humans and nonhumans in which roles, attributes, and interactions are exchanged. "If anything, the modern collective is the one in which the relations of humans and nonhumans are so intimate, the transactions so many, the mediations so convoluted, that there is no plausible sense in which artifact, corporate body, and subject can be distinguished."[10] Collectives, alternatively called sociotechnical assemblages, are what they are only when the nonhumans within them have been socialized, that is, given histories, flexibilities, and negotiational capacities to affect changes in the collective itself.[11]

On the surface, this notion of a collective, of sociotechnical assemblages, would thus seem to be close to the hybrid-cyborg figures of Haraway. In both cases the notions of subjects, artifacts, and society are blurred and hybridized. Distinctive subjects, isolatable artifacts, and varieties of associations are deconstructed. But underneath, there are very different programs at work. Latour's project, as he says, is "in fact to free science from politics" ... and in particular to show that there is an *"alternative to the myth of progress."*[12] In contrast, Haraway proclaims, "The cyborg is resolutely committed to partiality, irony, intimacy, and perversity. It is oppositional, utopian, and completely without innocence."[13] In the Latour case the aim is to free science from politics; in the Haraway case the aim is to politicize science. And both would seem to stand in contrast to the commonsensical stance taken by Pickering, who denies that there can be any easy substitutability between humans and their machines: "the idea that, say, human beings can be substituted for machines (and vice versa) seems to me a mistake. I find it hard to imagine any combination of naked human minds and bodies that could substitute for a telescope, never mind an electron microscope, or for a machine tool.... Semiotically, these things can be made equivalent; in practice they are not."[14] So, once again, whoever or whatever the nonhumans are, they have failed to produce a consensus among their defenders.

Yet the nonhumans remain, in their new context, actants in some sense. I shall now argue that there may be a robust middle ground upon which one area of consensus does emerge. I shall call this middle ground the area of interaction or performance. To open this new area, I shall revert to some of my own earlier notions concerning human-technology relations. In briefest form, I argued from the early seventies on that (a) the smallest unit possible for understanding humans and technologies (nonhumans) was the implied symbiosis of humans plus their artifacts in actional situations, (b) a phenomenological analysis of such interactions showed all technologies-in-use to be nonneutral by virtue of transforming situations in which such technologies were used, and (c) a *selectivity structure* (magnification/reduction) could be discerned as an invariant in such uses.[15]

What I am calling the middle ground gets striking confirmation from the uses Latour and I made of the same example—the denial of the NRA claim that "Guns don't kill people; people kill people." In *Technology and the Lifeworld* (1990), I claimed that my account was a relativistic one (in a physics metaphor):

> The... advantage of a relativistic account is to overcome the framework which debates about the presumed neutrality of technologies. Neutralist interpretations are invariably nonrelativistic. They hold, in effect, that technologies are things-in-themselves, isolated objects. Such an interpretation stands at the extreme opposite end of the reification position [of technologies—see Latour below]. Technologies-in-themselves are thought of as simply objects, like so many pieces of junk lying about. The gun of the bumper sticker clearly, by itself, does nothing; but in a relativistic account where the primitive unit is the human-technology relation, it becomes immediately obvious that the relations of human-gun (a human with a gun) to another object or another human is very different from the human without a gun. The human-gun relation transforms the situation from any similar situation of a human without a gun. At the levels of mega-technologies, it can be seen that the transformational effects will be similarly magnified.[16]

Thus I could not help but be struck when a colleague gave me a copy of Latour's 1993 paper (revised as chapter 6 in *Pandora's Hope*, 1999), "On Technical Mediation," in which this same example is more elaborately analyzed. Latour's context is precisely the same attack upon neutrality and reification noted above: "The myth of the Neutral Tool under complete human control and the myth of the Autonomous Destiny that no human can master are symmetrical."[17] Then, by granting actant status to both, Latour produces a complex analysis of how both gun and human are transformed:

> A third possibility is more commonly realized; the creation of a new goal that corresponds to neither agent's program of action.... I

called this uncertainty about goals translation. . . . I used translation to mean displacement, drift, invention, mediation, the creation of a link that did not exist before and that to some degree modifies the original two. Which of them, then, the gun or the citizen, is the *actor* in this situation? *Someone else* (a citizen-gun, a gun-citizen) . . . You are a different person with the gun in your hand.[18]

Beyond the obvious parallelism with my relativity context above, Latour goes on to claim full symmetry: "This translation is wholly symmetrical. You are different with a gun in your hand; the gun is different with you holding it. You are another subject because you hold the gun; the gun is another object because it has entered into a relationship with you."[19] Although from a framework of phenomenological interactivity, I would agree to the same conclusions about how subjects and objects are both transformed in relativistic situations; the disagreement would be secondary over whether or not subjects and objects are simply eliminated as meanings by virtue of symmetries. Here, then, is a middle ground regarding technological mediations.

With only slight modification, this same middle ground is accommodatable to Pickering's vocabulary as well. His naked mind, in the human-gun situation, enters into a tuning process and dance of agency as the human-with-intentions enters into the resistance and accommodation of machinic agency provided by the gun. The agency of the gun makes possible results that neither entity alone can enact— and does so in real time. But the transformations here remain short of full symmetry since only the human actant retains intentions. Symmetry is reasserted in Haraway's notion of the human-gun as cyborg. But, not without irony, the strict noninnocence (which is perhaps an intentionalist vestige correlated with technological nonneutrality) of this cyborg configuration is equally transparent.

My proposed middle-ground consensus gains the recognition that the varieties of technological mediations with their nonneutrality— and noninnocence—transform use-situations. And these entail some type of action, at least in the sense of an interaction, implicating the nonhumans. But I wish to suggest one final modification regarding the middle ground, which relates to the indirect means by which the nonhuman role may be discerned. I shall try to elucidate this by means of a narrative recasting the roles of the very humanists who have been accused of ignoring the nonhumans.

Maybe Latour reads the wrong humanists, or perhaps the humanists who lack humor and dress in black, according to Haraway, are those who leave the nonhumans out. Because when I read, say, Umberto Eco, I find novels full of nonhumans, fully socialized as Latour would have it, and full of actions, although located in the labyrinthine locations and esoteric practices of ancient libraries (*The Name of the Rose*), the chronological instruments that disorient (*The Island of the Day Before*), or the conspiracies of *Foucault's Pendulum*:

> That was when I saw the Pendulum.
>
> The sphere, hanging from a long wire set into the ceiling of the choir, swayed back and forth with isochronal majesty. I knew—but anyone could have sensed it in the magic of that serene breathing—that the period was governed by the square root of the length of the wire and by pi, that number which, however irrational to sublunar minds, through a higher rationality binds the circumference and diameter of all possible circles. . . . I also knew that a magnetic device centered in the floor beneath issued its command to a cylinder hidden in the heart of the sphere, thus assuring continual motion.[20]

Whether seen as constrained motion by Kuhn's Aristotle or as a pendulum by Kuhn's Galileo, this tale begins with the actions of a nonhuman. Eco fills his books with these nonhumans.

Rather than trace out the semiotics of humans and nonhumans by Eco, I shall tell a new tale, one disciplined by the attention to the nonhumans stimulated by science studies. I shall, however, locate the new tales in the precincts of those who presumably most ignore their nonhuman assemblages, the humanists. Moreover, since I wish to undertake this narrative in performative and interactional contexts, I shall attend to one set of core practices of active humanists, following them, as it were, to their own laboratories to observe them in action.

My first setting is an old one, premodern if you like. Our older, bearded humanist is standing at a sort of lectern on which he is writing on high-quality, cloth-content paper with a quill pen. The ink pot sits in a depression designed for it. Having reached middle age, perched on his nose is a set of spectacles through which he gazes at his productions. He is aware, since eyeglasses came into use in 1280, that men of his age could now continue with their reading and writing occupations whereas previously they had to desist at younger ages.[21] This "socialization" of eyeglasses has permitted a previously acquired practice to continue beyond its "time." Both the subject and the object have been translated by this extended practice. (The human-eyeglasses or eyeglasses-human parallel the example of the human-gun/gun-human as modified or translated collectives.)

Moreover, it is easy to locate a rather symmetrical interaction between the human and one of the nonhumans, the quill pen in this example. The human, acting through the pen as mediating technology, inscribes letters onto the paper; the pen both mediates and modifies the bodily action undertaken, and as the slot of the pen spreads or narrows with the handy motions of the writer, produces the letters of the beautiful script-character of premodern writing. Symmetrically, the pen is also modified by the subject writer and flexes—and gradually wears out—in the process. The object is observably modified, and the interactive transformations are quick to occur in real time. But it is less clear that the same symmetry applies to the other nonhuman relation of the context, the eyeglasses. The way in which the eyeglasses modify the vision of the writer is easily noted; his seeing is both transformed and made possible through the eyeglasses—but it is far less clear, at least in the use context, that the human user modifies the eyeglasses. No, the modifications that made the eyeglasses usable were made earlier, at the design and construction stage, with just such and such a magnificational power correlated to the stage of failure of aging eyes. In this case the symmetry seems less obvious, less symmetrical. Only by ascending to a much larger context, to too high an altitude, and then with too much generalization and abstraction, does symmetry emerge. (As previously, "there is no plausible sense in which artifact, corporate body, and subject can be distinguished.") Nonetheless, in the middle ground, interactive description of human-nonhuman relations, one can discern varieties of interaction and degrees of symmetry or asymmetry. The nonhuman actant is never fully transparent or pure but displays, albeit indirectly, its own role in the symbiotic context.

Leap now from the fourteenth century to the nineteenth century, and with it a change in the context of the humanists and, now, her nonhumans. The scene changes and we follow the writer into a small study with a desk and chair and a clumsy-looking writing machine atop the desk—typewriters were invented in the late nineteenth century. To preserve the analogue of humanist writing activity, we have her composing by typewriter; a letter or a novel is underway. Or perhaps she is Malvida von Maysenburg, one of Nietzsche's women friends, helping him with *The Genealogy of Morals*, commenting that "the poor man, almost blind, unable to read, unable to write (except on a machine)."[22] Nietzsche, after all, had noted, "Our writing instruments contribute to our thoughts,"[23] and typed these words out on a writing-machine (typewriter). But I rush this story, which is so filled with ironies. First, there was the designer fallacy that motivated the first typewriter designers. These were machines for those with poor eyesight or even those who were blind, machines to enable these persons to write. But the ultimate practices and collectives that formed were to be different—offices, bureaucracies, and agencies, staffed by the newly liberated young women who had replaced the male secretariat, were to be the fate of the typewriter. By 1881, an economist observed that "there are more women working at typing than at anything else."[24]

The males—who had previously been the secretaries—had, Luddite-like, rejected these new machines that no longer allowed the eye-hand coordination that formed letters, *belle lettre* style, which the pen had allowed. They could have employed Heidegger as their spokesperson, when he decried the typewriter:

> It is not by chance that modern man writes "with" the typewriter and "dictates"—the same word as "to invent creatively"—"into" the machine. This "history" of the kinds of writing is at the same time one of the major reasons for the increasing destruction of the world. The word no longer passes through the hand as it writes and acts authentically but through the mechanized pressure of the hand. The typewriter snatches script from the essential realm of the hand—and this means the hand is removed from the essential realm of the word.[25]

The typewriter of Nietzsche's time, of course, had not yet been improved into the speedy machines to follow, with proper timing built in by the QWERTY system that limited the speed of typists so they did not overtype letters. But the keyboards captured the preformed skills of young women, already familiar with the keyboard whereby their "piano fingers increased in economic value."[26] Nietzsche, however, probably appreciated the fact that two hands typing could also allow the flow of words to appear through the embodiment of body-machine-paper when he anticipated typewriter composition: "when my eyes prevent me from *learning* . . . I still will forge rhymes [by typewriter]."[27]

This new sociotechnical assemblage of humans and nonhumans, in which different writing instruments contribute to thought, displays the middle ground of interactivity in a different pattern than the first example. But it also displays its embeddedness in the wider social-izations that accompany it: gender role changes related to the tech-nologies, stylistic changes and more in the production of texts, a blur-ring of distinctions between composition and publication in print form. There is an implicit metaphysics of writing here that Nietzsche begins to discern.[28]

If the collective of humans and nonhumans changes by virtue of "writing instruments contribut[ing] to our thoughts" in different ways between pen and typewriter, a still more dramatic change is cur-rently happening with the contemporary and much more complex context of humans and nonhumans. The machinic agencies here are humans, computers, and the wider connections of the Internet. In both the earlier examples, writing remained a relatively simple process, a flow through machinic agency from writer's intentions into printed or inscribed form. True, the variables between pen and typewriter are quite distinct: speed of composition, difference in editing activity, in-clination toward different styles, appearance of the final text, etc. With word processing, associated with Internet capacities, writing is much more apparent as action-within-a-system.

Take two seemingly trivial, real-time variables as indices for writ-ing-as-system. When our human (let us imagine an adolescent doing homework) engages the word-processing program, the words do not immediately appear on the paper as they once did with pen or type-writer. Instead, they hover virtually on a screen, still soft and pliable, easily erased, corrected, spell-checked, modifiable. This tentativeness of the machinic inscription is so flexible that it posed a problem for Orthodox Judaism, which Michael Heim has pointed out:

> When the project to computerize the commentary on Jewish law got underway at Bar Ilan University . . . the programmers faced a puzzle. Jewish law prohibits the name of God once written from be-ing erased or the paper upon which it is written from being destroyed.

> Could the name of God be erased from the video screen, the disks, the tape? The rabbis pondered the programmers' question and finally ruled that these media were *not considered writing*.[29]

If the rabbis are right, the draft on the screen is not yet writing—it remains virtual "writing."

An inscription that is analogue to pen and typewriter appears only after one clicks "print" and the printer produces the edited copy on paper, with whatever format, font, or style decided, in real time. From the virtual draft to the printed draft, there has been the time and plastic virtuality for any sort of editing, which itself has been repositioned and forefronted by the ease with which it can be undertaken in word processing. But more, the program allows anything to be added to the text—pasted or downloaded text from the indefinitely large resource of the Internet, other files, e-mail, or material neither written by nor even virtually inscribed by the author himself or herself. The result displaces the singularity of the author-actor to some degree or, better, repositions the author-actor who is now in a role closer to that of a multimedia editor, pasting up a composite whole that will become the text. While the decisions as to what to download remain those of a quasi-deistic author, the product is one that entails actions of human(s), more or less distant, and many nonhumans in a result much more ambiguous than the lines of action taken in the previous two examples. The middle ground is collectively muddy! (I am tempted here to point to the parallelism that emerges between the presumed postmodern dissolution of subjects and authors and the sociotechnical practices that actually are followed in this embodiment of contemporary writing.) And if one then adds the complex system that is entailed with contemporary world publishing, the systems relations are even more complicated. (When I did my first book on philosophy of technology in 1979, my manuscript went to Holland for copyediting, on to Hong Kong for typesetting, back to the United States for proofreading, and then the whole process repeated, making for twice around the world before final publication of the result.)

Nonhumans on Middle Ground

My tale nears its end. If the demands of the nonhumans were not always heard, then through the therapies of contemporary science studies they have moved into a certain, although still disputed, visibility. My middle-ground claim is that there is, indeed, a limited set of senses by which the nonhumans are actants, at least in the ways in which *in interactions with them, humans and situations are transformed and translated.* I do not want to extrapolate this agreement too far, but rather argue that while there are some situations of clear symmetries, these are often limited, and there remain many situations that are asymmetrical. The objects (nonhumans) in such interactions modify the humans, the subjects are nonneutrally and noninnocently invariant, but the counterpart modifications are not always those of immediate, real-time modifications. Eyeglasses, typewriters, and computer systems change, are "improved," and provide different combinations of resistance and accommodation in different times. There are at best degrees and kinds of symmetries.

Certainly the nonhumans can no longer go ignored, but the degrees to which they can be socialized are, I suspect, both unclear and unpredictable. Part of their nature—still underestimated even in science studies—is the degree to which the unintended and unplanned results occur without intentions entirely. If we "dance" with the nonhumans, the steps that occur are often different from and often out of tune with the music played. That may be one of the reasons why, in the end, the cyborg metaphor of Haraway retains such suggestive power—of hybridization there is no end.

Bodies in the Philosophy of Technology

Part IV

Chapter 7
Prognostic Predicaments

Imagine a photograph of two adolescent Amish girls, clad in the traditional garments with bonnets and long, laced dark dresses, but wearing in-line skates while skating along a small-town sidewalk in Pennsylvania. This photo was publicized in newspapers and other media. For the larger public, the response is likely to be one of a sense of incongruity between the traditions of horse and buggy, nonuse of electricity, "plain clothing," and yet in-line skates—the latest, highest-tech skates, equally popular among the Lycra-clad, brassiere-showing urban youth in Central Park.

What the photograph does not reveal, however, is the process behind the Amish acceptance of this technology. As a colleague once pointed out to me, the Amish—in spite of the very technologically conservative approach taken by this religious community—have probably one of the most sophisticated and effective forms of technology assessment available. Every new technology is evaluated with strict considerations about whether or not it will support or enhance the values of the community, or detract from or erode those values. Thus, not everything—indeed not much—is accepted from the glut of current innovations that virtually immediately pervade the larger American or other industrially advanced societies. The in-line skate, it turns out, plays the same role for the Amish as the little red wagon, the scooter, or other forms of human-powered entertainment. It can fit, it would seem, into the plain lifestyle of the Amish community in spite of its apparent incongruity for the rest of us. Electricity, television and cinema, and most other high technologies remain excluded. I would only add that the technological conservatism of the Amish is not because of an antitechnological attitude, it is because of their deeply held (and conservative) religious beliefs.

I do not know if this community decision was a form of political compromise allowing Amish youth something new or not. It is known that many youth leave Amish society, yet the Amish have expanded into areas from Ohio to upstate New York, beyond their previous boundaries. The tensions among the Amish are not dissimilar to the breakup of Eastern European Socialist countries in face of the onslaught of Western technological entertainment and consumer technologies, which became desirable but unattainable or unaffordable in those societies.

The point of this vignette, however, is not to contrast hyperconsumerism and technological saturation with a religious form of minimalist nostalgia and communal values. I am using it, rather, as a hyperbolic indicator of a problem philosophers face with respect to changing technologies and their evaluation. The typical role many think the philosopher ought to follow is that of ethician or the reflector upon normative aspects of technologies within societies. How *ought* we to deal with technologies? What will their effects be?

I do this because both from within my own trajectory in the philosophy of technology, and increasingly from the recognition of other philosophers and historians of technology, there emerges a practical antinomy with respect to precisely the predictive problems in technological development.

The Philosopher's Prognostic Antinomy

The antinomy can be stated simply: if philosophers are to take any normative role concerning new technologies, they will find, from within the structure of technologies as such and compounded historically by unexpected uses and unintended consequences, that technologies virtually always exceed or veer away from intended design. How, then, can any normative or prognostic role be possible?

Philosophers, typically, are expected to play postdevelopment normative roles (as ethicians in applied ethics, for example). This usual role I shall call the Hemingway role. Ernest Hemingway's job in wartime was in the ambulance corps, and he reflected his experience in *Farewell to Arms*. He did get into the battlefield, he was actually wounded, but his task was to pick up the casualties. He was part of the battlefield cleanup squad.

This metaphor is appropriate for the many applied ethics roles occupied today by many philosophers. These began at first in the context of the development of medical therapeutic technologies. For example, during the early days of kidney dialysis, at first scarce and expensive, philosophers, theologians, and other nonmedical personnel were called upon as a "civilian ambulance corps" to deal with the ethical problems. Much of this relates to "lifeboat ethics" of scarcity situations and concerned decisions about who should get the limited treatment. The reason why the Hemingway role fits is that the Spanish Civil War ambulance corps, together with the nursing staff, had to practice triage on the spot: (a) who was possibly recoverable or likely to live? (b) who was borderline and questionable for recovery or life? (c) and who was dead or could not survive? Depending on the severity of the battle, the borderlines for triage could shift upward or downward.

I do not wish to discount the importance of the Hemingway role—or of applied ethics. These are clearly an improvement over the premodern form of cleanup process. After a medieval battle, and only after, the cleanup squad arrived. Then, sometimes after first stripping the dead and dying, the cleanup squad moved and cared for the injured. Not only was the chance for recovery lower, but the wounded had to remain on the field, bleeding, until the battle was well over.

But in both cases the metaphor points to the endgame role always played by the ambulance corps. The therapy and healing roles they enacted remained absent from the strategy rooms of the officers and military commanders, and farther still from the political considerations that always lie behind, before, and in the occurrence of war itself. For applied ethics in this context, it is always after the technologies are in place that the ambulance corps arrives.

I have argued on numerous occasions that if the philosopher is to play a more important role, it must be not only in or limited to the Hemingway role, but should take place in the equivalent of the officers' strategy meeting, before the battle takes shape. I will call this the R and D role.

A first response to this proposal might well be: but who wants any philosophers among the generals? the research and development team? the science policy boards? The implication is, of course, that philosophers will simply "gum up the works." And the excuse will be that philosophers are not technical experts, and any normative considerations this early will certainly slow things down—a sort of Amish effect. Of course, the objections in turn imply the continuance of a status quo among the technocrats, who remain free to develop anything whatsoever and free from reflective considerations.

It should therefore be noted initially that the antinomy I am pointing to arises primarily for R and D role philosophers. There is an advantage to be had from having to deal with already extant problems in the Hemingway role position.

But first, permit me to sharpen the antinomy. In my own work I have argued that *all technologies display ambiguous, multistable possibilities.* Contrarily, in both structure and history, technologies simply can't be reduced to *designed functions.* I have claimed that there is a designer fallacy that functions similarly to the intentional fallacy in literature. That is, if the meaning of a literary work cannot be traced or limited to the author's intent, similarly, in technology, its use, function, and effect cannot and often does not reduce to designed intent.

Heidegger's hammer is a simple example. A hammer is designed to do certain things—to drive nails into the shoemaker's shoe or into shingles on my shed, or to nail down a floor—but the design cannot prevent a hammer from becoming an objet d'art, a murder weapon, a paperweight, etc. Heidegger's insight was to have seen that an instrument *is what it does, and this in a context of assignments.* But he did not elaborate upon the multistable uses *any* technology can fall into with associated shifts in the complexes of "assignments" as well. No technology is one thing, nor is it incapable of belonging to multiple contexts.

The same obtains with complex technologies: e-mail in my university was first used to transmit memoranda, then as a substitute for phone tag, then for chain letters (which the administration tries hard to prevent), and even the propagation of computer viruses. Kittler has well shown that the typewriter (and one can add the telephone) was originally designed as a prosthetic device to help persons with sight deficiencies (or the telephone as a sort of hearing aid), uses that became, at most, marginal as the office soon transformed the secretariat through the typewriter and communications through the telephone.[1]

I argue that the very structure of technologies is multistable, with respect to uses, to cultural embeddedness, and to politics as well. Multistability is not the same as neutrality. Within multistability there lie *trajectories,* not just any trajectory, but partially determined trajectories. Optics takes us into the micro- and macroscopic as the histories of telescopes and microscopes evidences, but optics remains within the boundaries of the light spectrum and did not, by itself, develop into the new astronomy that now ranges from shortwave gamma rays to radio waves, thus revealing a wider world. Similarly, the externally fulfilled intentionality of seeing a Moon mountain carries with it not only the magnification of this external phenomenon, but it magnifies the motion of the observer holding the telescope and thus reflexively opens the way to a discovery of bodily micromotion—a trajectory not developed by Galileo but implicit within his favored instrument.

These complexities of multistability clearly make prognosis difficult, perhaps impossible if the aim is full prognosis. These are multiple intrinsic possibilities of the technologies. Historians of technology, however, tend to focus upon effects, and I will refer to two books that make the case brilliantly for the unforeseen and unintended uses, consequences, and side effects that all technologies produce.

The first book is *Why Things Bite Back: Technology and the Revenge of Unintended Consequences* (1996) by Edward Tenner. I do not have the time to outline in detail all the forms of a "revenge theory" of technological consequence—which include differences between *rearranging, repeating, recomplicating,* and *reconstructive* effects.[2] But the book is glutted with examples of each. His project began by reflecting upon a prediction made by futurologist Alvin Toffler concerning the coming of electronic media. Toffler says, "making paper copies of anything is a primitive use of [electronic] machines and violates their very spirit."[3] We are all aware of the "paperless" electronic society we now inhabit! Tenner claims, "Networking had actually multiplied paper use. When branches of Staples and OfficeMax opened near Princeton, the first items in the customer's view... were five-thousand sheet cases of paper for photocopiers, laser printers, and fax machines."[4]

I shall not even attempt to list all the multiple examples of these revenge effects—three simple ones illustrate how a single technology, which necessarily belongs to a context of assignments, produces unintended and often revenge side effects.

Cheaper security systems are flooding police with false alarms, half of them caused by user errors. In Philadelphia, only 3,000 of 157,000 calls from automatic security systems over three years were real; by diverting the full-time equivalent of fifty-eight police officers for useless calls, the systems may have promoted crime elsewhere.[5] (In my own village on Long Island, the situation is bad enough that the trustees are considering fines for each false alarm.)

Another example cited by Tenner comes from the well-known phenomenon where temperatures in cities are always higher than the countryside, due to pavement, stone, and concrete that retain heat and augmented by air conditioning, which shifts interior heat to the exterior. Air-conditioned subway cars spill heat upon the platforms, making them ten to fifteen degrees hotter, so that a ten-minute wait for a ten-minute ride actually produces a heat gain to the rider!

A final familiar example comes from the change of composition technologies—repetition strain syndrome and carpal tunnel syndrome were rarely known in the days of the typewriter, but have escalated with the computer. The harder and slower strike of the former yielded to the faster, lighter advance of the latter and contributes to this contemporary ailment.

We now have enough examples to clarify the antinomy. Technologies contain multiple possibilities for use, direction, and trajectory—are essentially multistable—making clear prediction of effect, use, and outcome difficult, if not impossible. And once in place, technologies produce, in the context of the multiple assignments to which they belong, unintended and often "revenge" effects, again difficult, if not impossible, to predict.

The second book I wish to cite is *Naked to the Bone* (1997) by Bettyann Kevles, a history of medical imaging.[6] Here, the unintended side effects arise from precisely what was discovered to be—only subsequently designed to be—a medical technology that finally revealed bodies as transparent. The history of the X-ray begins this process. Roentgen's discovery was publicized through his distribution of an X-ray photograph of his wife's hand with ring, showing bone structure under and inside the flesh. This technology, begun in 1896, was one of the fastest to acceptance of any process. But to obtain these early images, long exposures were necessary—up to seventy minutes in some cases. Retrospectively, we now know the result—severe radiation damage.

This knowledge did not occur immediately—indeed, one of the early uses of the X-ray was deliberate exposure to treat acne and skin disorders under long exposure time. By 1911, documented cases of burns, cancers, and even deaths had accumulated, and "At a meeting of radiologists in 1920, the menu featured chicken—a major faux pas because almost every one at the table was missing at least one hand

and could not cut the meat."[7] The history of the very instruments that make nonintervention possible for exploring the body, but which cause side effects by their use, continues to the present. This, too, is part of the unpredicted revenge effect.

This double-dimensioned prognostic problem is, I am arguing, more of a problem for persons playing roles in relation to prognosis—in our case, the R and D role philosopher.

Philosophers in the R and D Position

The antinomy clearly points to the difficulties of any normative, prognostic role. But before I make suggestions concerning how to lower these difficulties, let us take a look at a few historical examples of R and D philosophical attempts. Interestingly, the examples I will cite do not primarily belong to normative activity, but rather to epistemological aspects of technological development. They are, however, suggestive of a positive role for the R and D philosopher.

The most sustained example of the R and D role is the case of Hubert Dreyfus (he is not alone, but I shall use him as exemplar). In the early days of AI (artificial intelligence), Dreyfus was called in as a consultant by the RAND Corporation to analyze and critique the development of AI programs precisely because they were failing to deliver either as fast or as effectively as the proponents predicted. The result was an epistemologically scathing critique of the program, *Alchemy and Artful Intelligence* (1965), followed by several editions of *What Computers Can't Do* (1972, 1992).[8] At the core of the critique were epistemological considerations concerning how human bodies work in intelligent behavior. While many reacted to Dreyfus as an enemy, second- and third-generation computer designers began to see the alternative model Dreyfus proposed as positive (among these, Terry Winograd's ontological design programs in particular). And these results have now spread to a much wider front, evidenced in an article in *Science* on "The Space around Us," in which Italian neuroscientists have adopted "motor intentionalities" from Husserl and Merleau-Ponty into cognitive science.[9] This example is not only an exception to the applied ethics role usually expected, but an example of a philosophical insight being incorporated into both science and technology developments.

A second example comes from observations I have been able to make while commuting the last few years to northern European technical universities. I have come to know a number of philosophers located in these polytechnics—they are often lonely in the sense that there are no philosophy departments as such, although in some cases there are applied philosophy departments. These philosophers, however, often find themselves on interdisciplinary research teams and play precisely R and D roles.

My visiting role is frequently a secondary one: I am asked to review research proposals and give advice and criticism. Examples of such programs have included Herman the Bull, a genetically engineered bull with human genetic components designed to lower lactose allergies for humans who drink milk, and the hermeneutics of crisis in instrumental displays wherein reading multiple instruments itself may cause a crisis. Mixed or confused readings have often played roles in crises, for example, in the Three Mile Island incident and with complex medical instrumentation, which needs ergonomic design work. Here are philosophers (including indirectly me as consultant) engaged in situ at developmental stages. I applaud both these directions with respect to the R and D role I am advocating. But my examples are not primarily normative, and the prognostic aims are minimal.

Even these examples hide failures of prediction. Dreyfus, in effect, predicted that Big Blue could not have been developed, and contrary to my own expectations about "what can be done, will be done," Herman the Bull has been put to pasture "without issue," as the legal profession might put it. But we have now seen how philosophers have entered R and D positions.

Prognostic Pragmatics

The antinomy remains: both structurally and historically, technologies present us with multistable ambiguities that exceed the bounds of rational and even prudential prognosis. Yet to leave the situation there is not only to invite a laissez-faire technological politics, but to rule out even the possibility of critical reflection. I shall instead begin to outline a set of *prognostic pragmatics* that could serve, minimally, heuristic purposes.

If technologies embody, both structurally and historically, the possibilities of multiple uses and unintended side effects, and all instantiate these in particular fashions, then one exclusionary rule for prognosis can be advised: *avoid ideological (utopian and dystopian) conclusions.* A utopian version of this, cited by Tenner, is John von Neumann's 1955 "prediction of energy too cheap to meter by 1980."[10] A far less grandiose version occurred when philosopher of science Isaac Levi assured me that while he admitted that X-rays turned out to have harmful side effects, sonograms were bound to be totally harmless. Not more than a few months after this, I sent him a clipping about a study in Japan that indicated that frequently repeated sonograms seem to affect the central nervous systems of fetuses. Similarly, dystopian predictions include the worries of the nineteenth century over health effects of train travel—presumably so fast that it would cause heart problems. The prediction of side effects is not in itself dystopian, but pragmatically based upon long histories of similar side effects from all and any bodily intrusion—including nonradioactive ones. This is a generalized caution based both upon knowledge of the ambiguous structure of technologies and upon the related histories of similar instrumentation. In Kevles's history of medical imaging, it becomes clear that awareness of side effects has been amplified, and side effects are expected by today's practitioners. *No technologies are neutral, and all may be expected to have some negative (as well as positive) side effects.*

From within the expectation that there will be side effects, a pragmatic caution might be: *if any negative effects begin to appear, amplify these and investigate immediately; err on the side of early caution.* In the X-ray case, skin burns were recognized very early, but techniques in shortening exposure time were slow in coming. It was also known that lead shielding prevented X-ray penetration, but shields for technicians were slow in coming. Similarly, King James (of Bible fame) had already noted the noxious and negative health effects of tobacco in the 1600s—and we still do not have safety standards.

Technologies, unlike searches for theories of everything in science, thrive on alternative developments. *Enhance alternatives through multiple trajectories.* Here, energy production is a good negative example. R and D going into nonnuclear and nonfossil fuels has been scanty. In spite of this, solar development has become much more sophisticated and is finding wider uses—were R and D dollars deliberately directed toward a multisource base, we might find more promising outcomes. (In a forthcoming book, I will demonstrate how contemporary sciences have increased breakthrough discoveries by the deliberate development and use of multivariant instruments. This postmodern multiperspectivalism in instrumentation has implications for technologies as well.)

Design use experiments with nonexpert and different users. The unexpected uses—both negatively and positively—with the Internet are interesting in this context. Negatively, our son's soccer coach was caught in an Internet sting as a pedophile; positively, we found a rather idyllic isolated ranch run on solar and renewable energy resources through a travel page on the Internet. The Internet, interestingly, has displayed a respect in which dealing with technological prognosis is very like dealing with pornographic issues—that is, issues of freedom of expression, related to idiosyncratic attractants, make it extremely difficult to evaluate.

These heuristic suggestions are clearly not meant to be exhaustive. They are, at best, suggestive. They are more of a guide to what parameters to expect, rather than being specific particulars—but this problem is no worse or better than any other form of prognostic activity. They do imply that we need to have a deep insight into both technological structure and the history of technologies, which is best based on broad and interdisciplinary knowledge; that a critical take is called for, detracted neither by utopian nor dystopian aims; and that multiple variant approaches are likely to be the most promising for contemporary complexities. It is my suggestion that philosophers seek precisely those situations that allow the expansion of the R and D role.

Chapter 8
Phil-Tech Meets Eco-Phil
The Environment

Philosophy of technology (phil-tech), relatively new to the North American philosophical scene in the seventies, emerged from its largely European roots under a somewhat dark cloud of technophobic colors. The godfathers, Martin Heidegger, Jacques Ellul, and Herbert Marcuse as the most popular, portrayed technologies as Technology, a sort of transcendental dimension that posed a threat toward culture, created alienation, and even threatened a presumed essence of the human. Although this dystopic tendency was later moderated by younger American philosophers of technology who saw multiple possibilities for admittedly nonneutral technologies, the more pragmatic and more empirical effect took some time to mature.

Roughly concurrent, but somewhat later to arrive as a parallel and related subdiscipline, philosophy of the environment (eco-phil) began with what I take as a similar dystopic perspective. Early alternative technology worries over neo-Malthusian population explosions and unsustainable consumer practices were often at the core of the eco-philosophies.

I wish to undertake a pairing of phil-tech and eco-phil with respect to the worries often dominating these reflections on contemporary life. I wish to look at some more nuanced situations regarding utopic and dystopic tendencies, and suggest directions that might be developed in relation to the hopes that both fields might entertain.

Although I was not quite a charter member of the Society for Philosophy and Technology (SPT), I have been active in this group since the late seventies. And to set the tone for phil-tech, I want to recall my first entry into this group. I had been asked to participate on a panel responding to the work of Hans Jonas, whose work concerned the early days of biotechnology and related genetic and sometimes environmental studies. At first I did not know what issue to address—until I read the passages that related to Jonas's call for a new ethics, an *ethics of fear* as he called it, as the appropriate response to the biological manipulations that he perceived as threatening the very notion of a human essence. My role, then, became that of critic with an attack upon yet another Euro-American example of technological dystopianism.

I cite this example to illustrate what I take to be a deep set of intellectual habits that seem to be common to many both in environmental studies and in much philosophy of technology: congenital dystopianism. In the meditations to follow, I am going to look at a series of these intellectual habits that are commonly held between environmentalists, or ecologists, and philosophers of technology, and in each case give certain critical responses with the aim of redirecting concerns that should unite these two subdisciplinary matrices.

The Rhetoric of Alarm

Within the precincts of SPT, the best-known institutional group for philosophy of technology in North America, many commentators have noted the dominance of the dystopian. If there are godfathers of SPT, they have been Ellul, Heidegger, and the Karl Marx of industrial-capitalistic alienation. Every one of these godfathers, at least as interpreted within the SPT context, displays some variant upon the ways in which Technology has become the degrading metaphysics of late modernity and, insofar as environmental issues enter the scene, is taken to be, in industrial embodiments, the primary cause of environmental degradation.

But the same themes often dominate environmental precincts as well. In Lester Brown's introductory chapter to *The State of the World*, titled "The Acceleration of History," the now familiar refrain, borrowed and extrapolated from Thomas Malthus, is rehearsed: there are more humans on earth in the last couple of decades than in all previous human history (overpopulation), there is a dwindling food supply both in oceanic fishing resources and in agriculture (Malthusian arithmetic versus geometric extrapolations), and concerns for global pollution accelerate the degradation process.[1]

I am terming this the rhetoric of alarm. It is correlated to Jonas's ethics of fear and its purpose is—in Paul Revere fashion—to awaken the listener to the dire fate of a presumed environmental catastrophe, the late modern equivalent of the redcoats.

Historically, the rhetoric of alarm is the flip side of nineteenth-century progressivist utopianism. It seems hard to believe, but there were accounts in England that extolled the more brilliant sunsets of the Industrial Revolution, already known then to be caused by industrial haze in the atmosphere! Indeed, it may well have been that the utopian promises of industrialization and technologization at the turn of the century, by their very overextrapolation, led to part of the flip phenomenon in the mid–twentieth century. The promises of technological solutions to social problems, surpluses of food through agricultural revolution, democracy for everyone through communications, none of which occurred in either the time predicted or to the degree expected, may well have helped the distracters who linked technologies to everything from warfare to the Holocaust.

How, then, do we escape the horns of the utopian/dystopian dilemma?

Response: Critical and Skeptical Cautions

I am, of course, suggesting that excessive rhetorical strategies may work both ways. If promising too much can lead to disillusionment, then prophesying results that do not occur may lead to apathetic responses. Here my diagnostic will combine what I take to be the best habits found within scientific communities with those of a critical philosophy.

Excessive rhetorical strategies are often ill-founded and cause more harm than good. My first example for critical demythologization is the notion of a Malthusian extrapolation. In its original form, it was a hypothesis that populations would, in some time period, exceed the capacity of the Earth to feed them. In the long term, this could not occur if for no other reason than that excessive populations would cause starvation until a lower population level would return to some kind of homeostasis. Malthus himself eventually recognized this and modified his earlier theses.

Moreover, the contemporary version of such an extrapolation is even more ridiculous than its original Malthusian form. Statistically, the fact of the matter is that we had no way of accurately estimating human populations prior to a few centuries ago. This is not merely because we did not have census statistics for prehistory, but because—I contend—there is often a tendency to undercount, by virtue of another set of bad habits relating to historical foreshortening even within science, which oversimplifies prehistorical trends. Did you know, for example, that the precolonial populations of the Americas are currently estimated to be hundreds of times larger than the estimates of only three decades ago?

On a visit to Mexico, I got a look at the world's largest pyramid in Cholola. Its base is more than twice that of Cheops, but this pyramid was not even known until 1910 because it is literally buried under thousands of tons of dirt deliberately placed upon it. Surmising that if there is one such buried pyramid, there might be more, I entered into a conversation with a Mexican anthropologist and asked him if he knew of other buried pyramids. His response was that recent air surveys (using magnetometers that can detect undersurface structures) show that there may be 86,000 such pyramids in northern Mexico. In short, the probable population of northern Mexico in precolonial times just took a more than Malthusian leap in size. A similar population leap took place with the discovery in the last decade by similar processes of many cities of up to ten thousand people in the Mississippi Valley of the last millennium, one of which existed in Cahokia, Illinois. This city was larger than Copenhagen at the same time a millennium ago.

Another example relates to the so-called Neanderthals. Although recent discoveries show that there was a very widespread distribution of these near relatives of modern humans, everything we know about them is based upon only about thirty skulls, widely distributed. We simply do not know what population numbers to attach to this hominid species.

Robert Oppenheimer claimed, in the fifties, that there were more physicists alive then than in all pre-twentieth-century physics, to which Art Buchwald replied, "at that rate there will soon be two physicists for every person alive!" In short, a dose of good old-fashioned philosophical skepticism concerning historical accelerations may be warranted.

The same errors of scale occur on the future side of the extrapolation. As of 1997, the number of nations fallen below self-replication in population had risen alarmingly in the previous decade. While we have known for some time that industrialization is often followed by a population decrease—the average number of children per family in both the United States and Japan has shrunk from more than 5 prior to World War II to 2.2 or fewer since—the lowering of growth into negative numbers is still more recent. There is now evidence that similar falling birthrates are occurring in South American nations, most of which have fallen into a range of 2.1 to 2.8 children per family, not far from the 2.2 figure needed for a level population. Thus, were we to apply the negative reversal projection in Malthus/Oppenheimer/Buchwald fashion, might we not be justified in worrying about the self-reduction of the human population to zero by the end of the next millennium? Moral of the story: remain skeptical of extrapolative arguments that may underestimate pasts and overestimate futures.

Microsolutions to Macroproblems

An intellectual habit found in both philosophy of technology and ecological circles of philosophers, which has been applied, in my opinion, as badly as excessive rhetorical strategies, is the tendency to see problems as macroproblems but to propose microsolutions. Perhaps this is a variant upon "think globally, act locally."

Waste and recycling is an example of macroproblem/microsolution thinking. Waste and chemically toxic processes are clearly macroproblems, particularly acute in industrialized nations. Toxic waste (and toxic applications to agriculture, etc.) are the most extreme cases of this problem. What are the solutions? The most popular microsolution is recycling of waste, and this is accompanied by the usually ineffective solution of straightforward *banning* of toxic products.

I am not against recycling or bans—I am merely indicating that the microsolution by itself does not solve the macroproblem. Even if recycling were to rise in efficiency from its current 15 to 20 percent current effectiveness to some future 60 to 70 percent effectiveness, it would merely slow, not stop, the problem. Similarly, the banning of all toxic materials without substitute would not solve the problems of either cleaning up the environment or raising food productivity. It would, instead, displace the problems into different contexts since the bugs, microorganisms, and diseases that had previously been partially controlled would reassert themselves.

What I am pointing to is the tendency, the intellectual habit, to think "small is beautiful," which is, to my mind, equivalent to a form of nostalgic romanticism found among philosophers of technology and ecologists. For example, in the seventies the preferred solution to the globalization of technologies among the dominant groups of philosophers was to favor "appropriate technologies," i.e., small and simple technologies for (ignorant, untrained, and unready) Third World peoples. Given both the implicit romanticism and condescension involved, it is little wonder that such policy recommendations failed.

Today, that solution is currently replaced with the search for *sustainability*, a systems approach that seeks stasis (or, as one of my Dutch colleagues termed it, a new search for perpetual motion.) My experience of responses, when conferences on this issue are held, indicates that success in sustainability is as unlikely as was appropriate technology two decades ago. The reasons are not hard to discern, since so long as appropriateness and sustainability apply unequally to First and Third Worlds, there is likely to be little appreciation for the microsolution. As long as whole populations remain reluctant to deliberately lower living standards, the equalization process between First and Third Worlds will not take place.

Another, although more minoritarian, version of a microsolution for a macroproblem, again found within both philosophy of technology and environmental circles, is a form of *retrospective romanticism*. Here the shape taken is one that idealizes the microsolutions of indigenous peoples or of traditional cultures. One of our graduates, David Abram, won a major literary prize, the Lannon Literary Award, for his book *The Spell of the Sensuous*.[2] Abram draws from his experience of traditional cultures—mostly Native American but also many smaller Asian cultures—to argue that forms of respect for the environment and nondestructive practices still thrive among such peoples.

There are two problems with this retrospective romanticism. The first is that even where the microsolutions did work, they worked in contexts that largely no longer exist. Slash-and-burn agriculture can indeed work—if and only if the jungle area is large enough to recover between migrations of the agriculturists. But to maintain this situation, both territory and population size must be maintained. Populations that have increased, whether globally or locally, by virtue of the medical technologies that lower infant death rates and increase the number of adults who live into old age (never mind utopian wish fulfillment urges to longer lives), soon change this situation. Few romanticists actually advocate giving up all modern medicine! Nor do I find much appreciation for the massive bloodletting rituals of war and slavery that may have contributed to keeping overpopulation from occurring in Mesoamerica.

The second problem is that while some cultures did indeed attain human/environmental sustainability for very long periods of time (Australian Aboriginals, Arctic Inuit, European peasants), one should not overlook the many more societies whose premodern practices ended up in human-initiated environmental degradation (Mesopotamian irrigators, Easter Islanders, and, earlier, the humans who may have contributed to the extinction of the late Ice Age large mammals, or, more recently, our own civilizational ancestors who deforested the entire Mediterranean Basin). Not just any traditional culture or indigenous people can attain sustainability.

Response: Critical and Skeptical Considerations

What I am really arguing is that we have not yet fully diagnosed either what our technologies can or should do, or what the environmental crises are. So long as we naively accept both negative and positive hype—we will soon have ten billion humans on the earth, or we will soon be a chummy global village through the Internet—we are not likely to bring either technologies or ecosystems into appropriate focus.

Admittedly, to this point, I have exercised my own style of acerbic, somewhat cynical critique. Perhaps the time has come to turn to the positive side of the analysis and make some suggestions for how philosophers of technology and ecologists can begin to address the environmental problems we face.

Sizing Up the Problems

I suggested earlier that I would draw upon aspects of the scientific community at its best and its subsequent impact upon policy. Regarding problems of the environment, I first wish to point to the process and result of scientific consensus building. In today's global scientific communities, there are several recent consensuses. The planet is warming up as a whole, and there are signs of the beginnings of a greenhouse effect. This consensus, however, has been hotly debated and contested, even with respect to the degree to which homogenic causes contribute. But with the hyper-high-tech measurements we are now able to make, the scientific community has at least demonstrated that global temperature has risen, that ocean levels have risen, and that global weather patterns have changed in keeping with models of a greenhouse effect.

Within this consensus, hardly news to most of us today, one particularly recent and challenging set of observations relates to what could be called a middle-sized problem—the ozone hole. This phenomenon, only discovered within the last two decades (incidentally, several of the first describers of this phenomenon were Stony Brook earth science colleagues), is now known to be exacerbated by the rise in chlorofluorocarbon gases (CFCs), carbon monoxide, etc., many of which can be traced to particular industrial products.

This middle-sized problem, however, did not fall into the doomsday dystopian predictions, nor has it necessarily disappeared. Instead, it was recognized, publicized, discussed by the governments of the world, and with a relatively short political life span, got addressed through global legislation and agreements that appear to have already begun to slow or eliminate the accumulation of longlived greenhouse gases. The ozone hole has stabilized and even begun to shrink in the last two years.

There are several important aspects to note concerning this phenomenon. It was a clear indication of homogenic activity on a globalimpact level. We industrial humans partially caused it—but we also are underway in solving or reversing the problem. Another crucial element in the problem process relates to the building of consensus within science as a kind of intellectual motor to drive the corrective process.

The ozone hole is perhaps the most dramatic of late twentiethcentury middle-sized problems that have been addressed with some success. Smaller, more regional successes in the reversal of environmental degradation may also be pointed to. The Thames, which underwent cleanup processes helped by legislation in midcentury, now contains fish as far upstream as London, which had not been reported in over a century. Similarly, today there are signs posted along streams near my summer home in Weston, Vermont, showing how to identify the differences between trout and young salmon for fisherfolk. Salmon, just beginning to recolonize these streams, are still under protection, whereas trout are not so scarce or endangered.

Improvements in the amount of atmospheric lead in North America have occurred, as has the technology of more cleanly run combustion engines. One indicator is that the most recent move to transform two-cycle engines (outboard motors, lawnmowers, chain saws, etc.) shows how large engines (such as two-hundred-horsepower autos) are many times cleaner than two-horsepower chain saws. Even woodstoves, which used to be at most 35 percent efficient and put out much particulate matter, today reach 75 percent efficiency with little particulate matter.

Nor should we despair that such moves have little or no effect. There is an interesting evolutionary indicator concerning air quality. As early as the eighteenth century in England, and then in the nineteenth century in Michigan, the peppered moth began to turn from a very light-colored moth with few pepper marks to a much darker and highly peppered one in response to trees darkened by soot and industrial waste produced by factory processes. Today, these moths have returned to the light versions as trees turn lighter due to cleanups of the air.

All of this is at least an indication that reversibility can be attained in environmental areas. But I should like to point to a subtheme that is also important—the solutions I have been pointing to do not entail either abandoning technologies or pulling plugs. Instead, they often point to improved and higher technologies.

Air conditioning systems using CFC substitutes have been redesigned with higher efficiencies than the older machines that relied upon the greater evaporative and cooling properties of CFCs. Similarly, even presumably low-tech woodstoves have reverse draft systems that more efficiently combust oxidation gases. Our new Outback is peppier than my old Buick Roadmaster of college days, gets more than three times the mileage, uses lead-free gasoline, and has all-wheel-drive besides. In short, the solutions to technoenvironmental problems that have worked call for better technologies rather than older, simpler, or no technologies.

While I am far short of advising that high-tech solutions automatically solve the problems, I am suggesting that retroactive romantic returns to previous low-tech or simpler solutions sounds to me like a Bob Dole form of environmentalism. Take solutions from whence they come.

Changing the Technologies Changes the Problems

One of the most controversial biotechnological problems facing Americans is the deeply emotional one concerning abortion. Abortions are, of course, only indirectly environmental in that abortion can be one method of population control (as it is in Japan and Russia). Without examining the agonizing psychological and ethical problems associated with this process, one can note that the dominant form of therapeutic abortion in recent times has been the suction method used in the first trimester, usually performed as a sort of industrial version of biotechnology—the abortion clinic. Here, not unlike factory systems, the experts are gathered, time-motion studies for fast performance are entailed, and something like an assembly line occurs.

Again, following the industrial metaphor, highly visible factories have been easy targets for strikes, just as abortion clinics are the targets of abortion opponents. A change of biotechnologies, in this case the introduction of biochemical day after, or day after the day after, pills, when more fully implemented, will clearly change the way abortion can occur. A much more private client/doctor relationship, perhaps even followed by over-the-counter processes, will soon render the factory-strike metaphor, which advantaged protesters, obsolete and create a more decentralized, private occurrence, which will be much harder to make public and much more like miscarriages or spontaneous abortions, which are not condemned by abortion opponents.

None of the examples I have used are global as such. But I have shown how technoenvironmental changes are something of a "moving lunch" in which there can often be quick regional or problem changes. Similarly, the variants of "think globally, act locally" today are often quite ambiguous with respect to ecological issues. Reverting again to a New England example, today most New England states have much more forest than at any time in the last two centuries. Along with recovered forest land, long-absent species have returned (in the last two summers I have watched a wild turkey flock develop, and this year a young bull moose took a swim in my pond, both examples of wildlife not even present two decades ago), and local practices that disfavor clear-cutting remain in place. Maine, the last corporate holdout, has had to yield to a trimming back of property-holder rights as a result of the 1998 referendum. These recoveries, however, are often at cost to local preferences in that dairy farming continues to fail and local residents must convert to service industries (caretaking; lawn and meadow care; selective lumber arrangements; maple syrup mining, as it is called; and tourism industries), changing or degrading previous forms of rural or agricultural lifestyles. Nor have these changes, which today apparently favor both wilder forests and wildlife, totally eliminated climatological enemies such as acid rain (from Rust Belt areas—a far greater enemy to the environment than we landholding flatlanders), the warming change (which makes it harder for some species of trees to maintain health), or the invasion of televisual culture through minisatellites that threaten older lifestyles more than many other invasions. The ironic result is that the hardcore old-timer hunters from Vermont, despite statistics reporting abundant deer and moose, take their wilderness hunting to Canada and leave our environs to the flatlanders from Connecticut, New York, and New Jersey. Again, the technoenvironmental mix is one of high ambiguity and mixed results.

Moreover, these ambiguities also call into question the usual folk wisdom of many in both philosophy of technology and environmental philosophy, at least with respect to the advice that arises from the various forms of nostalgic, retroactive, or smaller-is-better romanticisms. I have been suggesting that there can be higher-tech solutions to lower-tech problem contributions. I have been suggesting that not all local, regional, or traditional solutions either solve the problems or are necessarily the best solutions. Above all, I am suggesting that technoenvironmental problems in the late twentieth century are almost all complex, interrelated, ambiguous problems that rarely yield to quick-fix or easily grasped solutions.

Critical Realism and Technoenvironmental Problems

It is now time to take a concluding look at the hard problems.

I have not meant, in any way, to distract us from facing what I have elsewhere (in two books, at least)[3] called a foundational problem for the philosophy of technology, that is, the solution to environmental problems. What I have intended is to point out that solutions to these problems must take shape at the appropriate levels of complexity and in contemporary contexts of technologies. In this conclusion, I shall look not at how we can "win the war," because I do not have any quick fix for that long-run strategy, but instead at some tactical areas where small battles have been won and at the placement of personnel and developments that may suggest tactics that work pragmatically.

I return to recycling, a presumed local solution that addresses part of a problem. Ironically, recycling is not local. Indeed, massive recycling is performed in China, which is the recipient of much U.S. and Japanese junk. The Chinese have complained that the Americans, in particular, are sloppy with their separations—too much garbage gets into the material. Recycling, like much else in the late modern world, is a multinational process. There is little, if anything, that is limited to the local.

Our industries ship goods to lower-cost labor areas to construct, so we send our recycling to lower-cost hand-labor areas for separation and processing. But the high-tech Germans may come to the rescue with a high-tech solution to the Chinese complaint. They have developed a version of the old cream separator for recycling that will eliminate the hand labor. (Turn the crank and the high-speed centrifuge separated the milk from the cream.) This device simply takes all recycling products and melts and crushes them down into a semiliquid. Through heat and other processes, the resulting paper residue comes out one spout, glass another, plastics yet another (Luddites, prepare to protest de-skilling!)

This process, indirectly, points up the hardest nut to crack—how does one turn multinational corporations green? Again, I have no quick fix, but in a few cases there are actually some positive indicators. At a conference, I heard a paper on how several high-tech companies have taken green directions. Both Xerox and Canon were stimulated by studies within the companies on recycling processes (of toner cartridges) that actually saved the companies money. These promising initial processes led to larger changes. For example, the old process of shipping machines entailed large cardboard containers, volumes of Styrofoam peanuts, and wooden pallets, few of which were recycled. A change to transparent thermal heat wrap, with a deposit return on the pallet, has not only saved hundreds of dollars but had an interesting positive side effect. Handlers, actually seeing the fragile machines, seemed to take more care than in the older cardboard box days, and shipping damages went down. The point of this example is that when green processes can be demonstrated to produce lower costs or contribute to higher profits, corporations will adapt accordingly. It takes innovators and pushers to develop this green efficiency. I am, however, convinced that green high-tech processes can do precisely this.

What role does or can the philosopher play in this green turn? For a long time, I have argued that our usual role is one doomed to *reactive* status. This is particularly true in the case of applied ethics, which began with medical contexts and spread to business schools, law, and other contexts, where the presumption is that technologies dispose, and it is up to us to make that disposition as ethical as possible. I do not want to unemploy all those philosophers who have made livings doing just that, but I prefer a more *proactive* position.

The proactive position that I am advocating places the philosopher at ground level, particularly in the *research and development* phase of technological processes. For a while I despaired of this happening, but in recent years I have observed and even participated in several of these projects. Ironically, some of these occurrences have happened precisely because of the displacement from traditional (Germanic?) academic vocational roles. In the United States, for example, some philosophers, lacking full-time academic placement, have found themselves freelancing part-time in computer companies or other high-tech companies (several of our alumni have found such positions). Here the kind of disciplinary thinking for which we have been trained comes into play in a new situation. Critical thinking can be applied to developmental, not resultant, processes. Contributions to some of the green directions just mentioned have been helped along by precisely such part-time philosophers.

More interestingly, in the last few years I have been doing quite a bit of commuting to northern European technical universities (Scandinavia, Holland, northern Germany), where I have found increasing numbers of retooled philosophers, particularly philosophers of science, working not in traditional departments of philosophy, but within these polytechnics. They often become part of interdisciplinary research teams of engineers or other applied sciences. These kinds of situations are challenging and cast a very different perspective upon philosophical application. I am suggesting that, at the very least, this front-loaded position should be one for *engagement* just as much as the end-loaded situation our ethicists have found themselves within.

Finally, I do not want to conclude with any easy sense that all technoenvironmental problems are simply fixable by applying more technology or that the removal of technologies does not solve the problems we now have. I want to indicate that all technoenviromental problems are complex, ambiguous, and interwoven. The tasks are not easy, but neither utopian nor dystopian attitudes ultimately help. It is precisely within the ambiguities and complexities of our current situation that philosophers must take their places, and it should be obvious that technoenvironmental problems take their very shape at these locations. What better place to be at the beginning of the twenty-first century?

Epilogue
Technoscience and "Constructed Perceptions"

The focus of the previous chapters falls upon embodiment and perception in relation to technologies. Although I have referred to many kinds of embodiment-related technologies, including ancient ones, I have pointed to the emergence of the complex, compound contemporary technologies that involve *virtuality, simulation,* and *computer modeling* (including *tomography*). These technologies—and the hype that goes with them—have special implications for embodiment and perception.

I have also contended that there is a convergence between virtual, simulation, and computer-modeled (hereafter VSC) technologies within science practice and popular (especially entertainment) culture. In this epilogue, I want to make some observations about this convergence and draw some consequences about it. The issues are complex and often confused, in part because there is overlap and loose usage concerning all the VSC technologies. What, for example, is the virtual? In popular form, one can enter virtual reality in a virtual reality arcade. The two most popular game versions include the donning of "face sucker" masks with tiny video screens close to the eyes and gloves that are wired to provide tactile-kinesthetic sensations, or entering a simulator that is enclosed, a sit-down "vehicle" with graphic audiovisual effects and possibly also movement of the simulator itself.

Unfortunately, the description and claims made for this arcade experience are usually cast in the now antiquated frame of early modern epistemology (Cartesian or seventeenth century). One is supposed to think of the virtual as a highly sophisticated and enhanced form of representation that duplicates reality. In this Cartesian epistemological sense, the old evil genius trick could (and does) reemerge: could I be fooled into thinking that virtual reality is the same as real reality? But in postmodern hype, the ante is raised so that the question becomes: is virtual reality *better than* and *substitutable for* real reality? I reject this way of framing the issue precisely because it retains this now outdated seventeenth-century epistemology that does not recognize embodiment or performance or the production of knowledge. Instead, let us look at some of the varieties of virtuality that are technologically produced and take account of what they are and do in terms of technology-human relations.

Some virtual reality technologies relate to a wider set of bodily-sensory dimensional experience than older technologies. This kind of VR technology adds some aspect of tactile and/or kinesthetic effect to already standard audiovisual effects. Presently it is probably fair to observe that audiovisual technologies are the implicit norm for many of the imaging, communicational, and entertainment productions of today. Cinema, television, some aspects of the Internet, teleconferencing, and on and on, are audiovisual technologies. Here, experienced relations are those that partially engage the perceptual embodiments of humans on both an individual and a social scale. We pretty much take the audiovisual norm for granted. The newer, virtual extrapolation—still largely primitive—adds partial tactile-kinesthetic experience to the audiovisual. This may range from minimal hand-operated interactivity (a joystick, for example) to fully wired bodysuits for higher degrees of virtuality. But if one takes nontechnological, active, whole-body experience as the norm, one can easily see that even these technologies remain short of RL. Even the best virtual reality machines fall short of the full and ordinary richness of bodily experience, and if one describes this in terms of the equally antiquated version of five senses, there remain lacking not only olfactory and gustatory virtual reality, but also full ranges of bodily activity. Moreover, to experience this enhanced virtuality of touch and motion, one must first enter a highly framed context. It is a theaterlike situation wherein one enters with a suspension of belief. Motility is also highly constrained—the player who walked off the VR platform would pull the plug.

At first, it would appear that there is little convergence between this style of the entertainment version of virtuality and science practice—but there are some significant exceptions. As I have indicated, science practice in the ordinary production of its evidence is even more reductive of whole-body experience than its entertainment counterpart, insofar as it relies upon a highly sophisticated visual hermeneutic for most of its results. If one wants to know in a medical context if the patient has a brain tumor, the production is one of a visual image by means of MRI, PET, or CAT scans or all three combined by computer tomography into a single, compound image. One then sees/reads this visual product with the trained, critical hermeneutic vision of the expert.

But in other areas of practice there are significant and ever more frequent uses of virtual processes. Suppose, in the context of a laboratory dealing with nuclear materials, our technicians must take a vial of radioactive material and move it from one place to another. One set of technologies in this task is robotic claspers (artificial hands, one might say) that can be manipulated to clasp the vial, lift it, and remove it to the designated new location. The human user manipulates a two-handed device that is connected to the remote robotic arms. For this to work in a human-user, technologically mediated movement, the robotic arm must supply kinesthetic-tactile feedback that the technician can "feel" so as not to crush the vial with too much pressure or to drop it with too little. This is a version of the same virtual or multidimensional technology operating in a nongame context. Pair this with remote-sensing devices—the Sojourner robot on Mars—and one begins to see some of the applications of this convergence. In the Mars example, there is not only a lack of kinesthetic feedback, but also a significant transmission time delay so that the results of some commands were crashes into target rocks. There remains, of course, the difference between user communities: in the arcade, virtual reality technologies produce a fiction for entertainment, whereas the robotic device, near or distant, undertakes a task within a real environment. But both are utilizing virtuality technologies.

Virtuality in the above sense overlaps with simulation technologies that can have the same degrees of multidimensionality. I have referred to early flight simulators, such as the Link Trainer. Contemporary flight simulation employs much more sophisticated virtual reality technologies—sound and motion are included, and the realism or hyperrealism of the projected landing or crash scenes has the latest three-dimensional dynamics, with real-time technologies that simulate bodily temporality. (The use of three-dimensionality is also often called virtual.) What simulation technologies entail is a situation-context. The participant, or player, enters a situation, and the degree of realism often includes random or unpredictable scenarios that build in more lifelike elements. Yet in the end, the player-participant, unlike the fictionally naive figures projected in science-fiction fantasies, always is aware that he or she is in a quasi-fictional context. One set of important differences between the arcade simulation and the annual simulation test for pilots lies in the social embeddedness of the two "games." The arcade player has little to lose other than ego or pride in the final score; the pilot who fails or displays lack of discipline can lose much more. Once again, the community of users structures the same technological context differently.

Given the shortcomings of fully tactile-kinesthetic technologies, there is a contemporary hybrid virtual situation that occurs. In this case, the combination of largely visual (or audiovisual) technologies with interactive hand controls produces the Nintendo phenomenon of action at a distance. *The Last Starfighter* is a science-fiction movie in which an Earth boy who had become the quickest gun in video-game simulators is transported to a distant galaxy in which one side in a desperate war needs a fighter pilot to save the threatened good guys. He becomes the hero, able to outgun the enemy pilots by virtue of his skills. What is interesting here is that the reduced set of bodily actions in eye-hand coordination for quick and precise results becomes, in a different context, another indication of pop culture/science convergence. Edward Tenner points to the emergence of "Nintendo surgery." Laparoscopy and other increasingly minute fiber-optic-assisted procedures attempt to make the best of the currently limited possibilities of virtuality in surgery. Visualization occurs through the instrument although distant tactile-kinesthetic development, Tenner contends, is poor:

> A television image is narrower, grainier, and of course flatter.... Surgeons can't feel internal organs with their fingers. To perform what critics have called "Nintendo surgery" requires skills different from those for traditional procedures.... Will a future generation of surgeons, growing up with video games and other flattened mappings of the world, learn a new way of relating to the body, as their predecessors learned to interpret stethoscope sounds, microscopic images, and X-rays?[1]

(Tenner considerably underestimates the resolution of some "Nintendo" practices. When I underwent a colonoscopy, fully awake because I wanted to see the imagery, the glowing, full-color screen showed in great detail the terminus of the scope with its tiny cutting apparatus—magnified on the screen—ready to snip any polyps to be found. While this was Nintendo in the sense that it was eye/hand for the operator, it was quite clear and vivid in visual resolution.)

This is an interesting convergence within the capacities provided by the limited current stage of this style of virtuality technology. At a surface level, such a convergence might seem to be the unsurprising outcome of the fact that many different communities use similar technologies and techniques. Astronomers and television producers both use enhancement, contrast, and digital manipulations to produce the images and results that they show. Since such complicated technologies as computer tomography, digital processes in both visual and audio technologies, and simulation technologies can all be used, a convergence of such uses seems a normal outcome. But this normality also flirts with old notions of *technological determinism*, and one must be careful not to revive this simplistic explanation.

Technologies do not determine directions in any hard sense. As some of the previous chapters have argued, while humans using technologies enter into interactive situations whenever they use even the simplest technology—and thus humans use and are used by that technology, and all such relations are interactive—the possible uses are always ambiguous and multistable. No designer can build in some single purpose or use, and thus there is no clear unidirectional determinability to even the simplest example. The simplest identifiable tasks, technologically, have multiple possible solutions. How does one eat soup? Many different cultures have separately invented spoons to perform this task, but many other cultures drink soup directly from the bowl, and some combine drinking from a bowl with the use of chopsticks for the bits. How does one fire a bow in archery? Some use a pinch technique, others use full-finger draws, and today's high-tech versions use a triggerlike pull. Moreover, some users hold the bow at a distance and pull the drawstring toward the body (English longbow and most Western users), while others keep the drawstring positioned and push the bow away (Mongolian horse archers and other Asian techniques—these also show that the very term, "drawstring," is a Western term). In both these simple examples there are, of course, some differences in the artifacts—chopsticks versus cutlery, long ash bows versus short composite bows—and in the context of uses. One lifts the bowl versus leaving it on the table, one stands at attention to fire versus riding a horse at a gallop, etc. Yet while the multistability is obvious, it is not totally indefinite. It shows a similarity of pattern and often what I have called a technological inclination, which approximates a possible "soft" determinism or direction. Eating with utensils does not totally, but it may dominantly, displace eating only with one's fingers; hunting or fighting with bows does not displace hand-to-hand hunting or fighting, but does transform the distances at which kills can be made.

What occurs with the simple examples applies with more complexity to the situation facing us with the high-tech examples considered here. The convergence I have noted is one in which VSC technologies come to the fore. Combined, these technologies, perhaps more than with previous technologies, enhance the sense of *constructed results*. The term "constructed perceptions" in the title hints at the postmodernity of the situation. If perceptions are what a subject has when sensing something, how could these be constructed? In English, however, there is a second common meaning as well—a "perception" is also taken as a belief of sorts and is usually contrasted with some notion of reality. Such *socially dimensioned* perceptions do, of course, have real effects.

I live in a region of Long Island that has had and is now undergoing a series of major controversies about the social and political impacts of technoscience development. One of the major issues of the not-too-distant past had to do with the proposed opening of a five-billion-dollar nuclear plant, Shoreham. In the end, public perception concerning its potential danger in case of a meltdown led to political movements that caused its closure (not too long after the Three Mile Island event in nearby Pennsylvania, but well before Chernobyl). It was sold to the state for one dollar and is now what one could call a technology monument—empty and unused. Today, a similar controversy rages with respect to the Brookhaven National Laboratory (BNL) and its nuclear research programs. A leak in a storage container (and several other sources) showed up as measurable radioactive traces in an underground water plume and in local rivers stemming from the BNL location. Again, there is a public perception about the possible danger of long-term, if low-level, radioactivity, which, in turn, has political effect. Several of the reactor programs have already been closed.

My colleagues and I in the Science Studies Forum and the Techno-science Research Group are involved with a number of aspects of these controversies, and in particular with the problems of trying to understand the hermeneutics of public/science relations.[2] But the significance of this version of perception that I am after is this: "perception" in this context is indirectly related to what I have earlier called body two perception, the dimension of perception that is situated within a sociocultural matrix that informs and can permeate body one perception, which is bodily and sensory. While I do not believe one can reduce bodily-sensory perceivability to its solely social dimensions (presumably this is what some take to be the core of social constructionism), neither do I think there is any bodily-sensory perception that is without its sociocultural dimension. Rather, these dimensions interrelate something like a core/field or figure/ground gestalt within the relation between embodiment and environment or world. What makes the situation even more difficult to analyze relates to the indirect ways in which body two or cultural perception occurs. I shall try to illustrate that while moving simultaneously closer to the roles of VCS technologies.

Hyperreality and Hype

In 1999, a massive global phenomenon occurred precisely because of the effectiveness of globally connected technologies, the Y2K phenomenon. The convergence of a calendar millennium (which in the West carried many quasi-religious undertones) and the global interconnectedness of computerization and computer culture became situated by a media-conveyed global hype. Ranging from social hysteria, evidenced by the many Y2K cult developments, to survivalists, to media-hyped pronouncements, many were convinced there would be major social and technological breakdowns due to computer-linked 1999/2000 turnovers.

This social perception had actual effects, and governments, primarily ours, spent billions of real dollars for preventive actions and programs. Yet when the actual turnover came, there were very few technologically caused problems, and to highlight the hype phenomenon, no more disruption in unprepared areas of the world compared to fully prepared areas. The jokes about tag sales for survivalist surpluses were widespread, while the technicians and retooled programmers could laugh all the way to the bank this time.

Computer as Epistemology Engine

Within and alongside all VSC technologies is the most important technology, the computer, or rather computers and their networks. What modeling, tomography, and various digital processes have made available is an *interchangeable and reversible set of exchanges between data and words, numbers and images.* For example, the traditional mode for quantitative analysis usually entailed some version of reduction of a phenomenon to quantities or digits. One sought to find a formula or algorithm for, say, a geometrical formation. Computation through computers can reverse that process; fractals and so-called chaos phenomena are popular examples of such reversibilities. By programming in a formula or algorithm, one can move from digital quantities to imagery. The surprise was that the visual results quickly were seen as real or naturelike. Fractal images could look like coastlines, mountains, or desert landscapes because of their complex and reiterative patterns.

This is an example of modeling that today is extended and extrapolated into many other phenomena heretofore too complex to begin to analyze, such as weather patterns. But the transformations of data into gestalt images, and vice versa, are processes that occur in both the sciences and popular culture. The image/data interchangeability occurs in most NASA programs. Interplanetary probes scan planets of our solar system with a variety of instruments ranging from optical through radar to magnetic scans. Results are reduced to transmissible data, transmitted, and then reassembled as images.

This translatability, however, also implies the possibility of manipulation, and through it a type of construction now made easier through computer processes. The movie *A Perfect Storm* has a scene in which one of the actors hangs suspended from a failing line on his fishing boat, and in the background we see an impossibly large wave about to engulf him and boat. This effect was created by photographing the actor hanging from the boat prop in the studio against a blue background, onto which the digitally produced impossible wave was retroprojected. The result is the dramatic shot desired by the movie maker. The introduction of digitally produced computer imaging in

Jurassic Park or in animated form in *Chicken Run* is becoming commonplace in film. Exaggeration and impossible effects are now normal.

The same possibilities occur for science imaging. Edward Tufte has pointed out an excellent example concerning the images that stimulated the creation of the Flat Venus Society. He shows a published version of images of the mountains of Venus derived from the 1992 Magellan flyover, in which the vertical scale is increased by a factor of 22.5.[3] To the credit of astroscientists, the Flat Venus Society was formed to protest this exaggeration, since there are no known mountains with more than a 3 percent grade on Venus. But I was surprised to discover that these distorted images appear, nevertheless, in respectable books on astronomy, without mention of the vertical exaggeration.[4] Of even more interest is the relation of science imaging to a background of political and funding interests. A series of asteroid images that appeared in *Science,* some with enhanced coloring (an example of enhancement and contrast used both in science and art), included a digitalized view of Earth from just behind the asteroid—as if to suggest a close flyby or collision. Possible connections to anti-asteroid programs are not hard to detect.

VSC technologies allow and facilitate manipulation and construction. In some cases this constructability simply makes previously difficult or impossible phenomena visible, measurable, and evident. Detailed three-dimensional modeling of the interiors of early hominid crania, for example, shows this possibility.

Such VSC technologies can accomplish what all previous instrumental realism technologies aim at—the bringing into perceivability that which was not previously perceivable. The phenomenon is brought within range of human embodied activity. Yet at the same time, precisely because this is a production, the viewer cannot simply take the newly constructed perception for granted. It may be both true and constructed, but it is constructed, and this calls for a critical take upon what I have previously called second sight.

In science, while the result is usually some visual display that can, as with all perceivable phenomena, be taken in at a glance, the production that lies within and behind the display is not always visible. For example, three different surveying processes were used to produce an apparently simple and accurate three-dimensional map of an ocean floor. The technologies employed used satellite passes, multidimensional sonar scans, and seafloor photography, all combined by computer tomography processes. The map, which was needed by the major contestants in the Cold War era to find places where submarines could hide or be found, was constructed by a series of satellite passes that mapped bumps on the ocean surface that, in turn, by inferential calculations entailing gravity effects, revealed sea mounts below the surface. Then multibeam, side-scanning sonar was used to fill in more resolution through ship-towed, underwater instruments. Finally, deep-diving photographic probes were used for the areas calling for the highest resolution to fill in the gaps. The multiple instrument scans are combined by computer processes, and the result is an apparently simple three-dimensional map. This multiple instrument/ computer-synthesized production of imagery is today a standard for such complex phenomena.[5]

The irony of contemporary high-technology instrumentation is this: scientific instruments can more clearly, more precisely, and more profoundly deliver data/images than in any previous era of human history. But they do this by means of higher degrees of active construction, intervention, and transformation than at any time in human history as well. There is something of an inverse proportion law at play—the better the data/image, the more constructed it has been. And in most cases, these technologies include VSC technologies. Virtuality, whether entailing wider ranges of bodily-perceptual experience or merely heightened (constructed) perceptual presentations such as three-dimensional imaging; simulation, whether in terms of bodily interactions through distance or simulation technologies, through computer models of such phenomena as global weather patterns, or through tomological and topographical composite constructions such as in medical (MRI, PET, CAT) technologies; or the compounded processes used for various mapping results, all bring higher and higher degrees of constructed perceivable presentations to the human observer.

Similar VSC technologies are at play in popular culture and its entertainment contexts. Virtual and simulation technologies may be enjoyed in the restricted frameworks of the arcades, and effects in science-fiction and other highly exaggerated catastrophe movies are the norm of today's viewing. Many times this leads to the conscious or subconscious awareness that what is experienced is simultaneously real and yet constructed. This conviction concerning construction plus some kind of reality leaves standing the very question I alluded to in my first imaging seminar. There are vast differences between the communities of image users with respect to what they hold they have constructed and of what reality is implied. It is in and between these differences that the critical arguments take shape. Science continues to argue that it presents the instrumentally real; entertainment argues that it presents the fictionally real of VR, which carries the utopian promise that it will be better than RL. But while the postmodern hype is such that it emphasizes the direction, RL toward VR, the opposite direction is not only possible, but, I contend, more dominant. That is the movement from the fantasy of VR back into RL. My example in this case is not from science practice, but from everyday life. It is another tale about Mark.

Last summer I began to teach Mark how to drive. My vehicle is a 4 × 4, manual transmission, Toyota pickup truck. The scene is the back road of our Vermont summer home. To embody driving, Mark must learn to coordinate the four gears with clutch and accelerator movements, and, as with all beginners, the first attempts are jerky starts, sometimes the engine stalls, and other normal first shakiness. But soon the starts become smoother, the next attempts move into higher gears and hence higher speeds, and there is the teenage exclamation, "Awesome!" After several such lessons, Mark turns to me and says, "You know, Dad, none of those simulator games really prepare you for *this!*" I took that as the telling indicator of the move from VR to RL, the move from the theater to the ordinary world that still exists, albeit alongside and interpenetrating the lifeworld.

Bodies in Technology

In the contexts of each conversation about bodies, with philosophy of science, with science studies, and with philosophy of technology, I have tried to forefront the role of embodiment and its perceptual expression. Embodiment, however, is always relativistic in the sense that it is a *relation* between the human and the technologies employed. What stands out first is that all human-technology relations are two-way relations. Insofar as I use or employ a technology, I am used by and employed by that technology as well. This is what Andrew Pickering calls the dance of agency and what Bruno Latour terms the symmetry between humans and nonhumans. In the second place, through our various journeys it can be seen that bodies, our bodies, adapt to different kinds of technologies and technological contexts. The Nintendo phenomenon that emphasizes eye/hand actions has been seen to span bodies in technologies ranging from video games to surgery and is a new, if restricted, style of movement that is very far from bodily sports activity or dance, whether classical ballet or modern. This range of adaptation to our machines, however, is not infinite or totally malleable. It reaches limits and has structural aspects, and this is one of the secret motives for the sophisticated attempts to bring science imaging into human perceptual range and for the virtual developments of that same direction to simulate dynamic three-dimensional effects. This points to the third aspect of bodies in technologies—the technologies must also adapt to us. A scientific instrument that did not or could not *translate* what it comes in contact with back into humanly understandable or perceivable range would be worthless. It would lack the anthropological invariant that points to the implied limits of the machines we build and use.

We are our bodies—but in that very basic notion one also discovers that our bodies have an amazing plasticity and polymorphism that is often brought out precisely in our relations with technologies. We are bodies in technologies.

Notes

Introduction

1. For the result, see "The Desire to Be Wired" by Gareth Branwyn, *Wired* (September/October 1993): 62–65.

2. Marco Ciannochi, *Leonardo da Vinci's Machines* (Florence: Becocci Editore, 1988), 12.

3. Ibid., 13.

4. Paul R. Gross and Norman Levitt, *Higher Superstition: The Academic Left and Its Quarrels with Science* (Baltimore: Johns Hopkins University Press, 1997); Noretta Koertge, ed., *A House Built on Sand: Exposing Postmodernist Myths about Science* (Oxford: Oxford University Press, 1998).

5. Steven Shapin and Simon Schaffer, *Leviathan and the Air-Pump: Hobbes, Boyle, and the Experimental Life* (Princeton, N.J.: Princeton University Press, 1985).

1. Bodies, Virtual Bodies, and Technology

1. Maurice Merleau-Ponty, *Phenomenology of Perception*, trans. Colin Smith (New York: Humanities Press, 1962).

2. Martin Heidegger, *Being and Time*, trans. John Macquarrie and Edward Robinson (New York: Harper and Row, 1962), sections 15 and 16, 95–107.

3. Don Ihde, *Technology and the Lifeworld* (Bloomington: Indiana University Press, 1990), 75–76.

2. The Tall and the Short of It

1. Susan Bordo, "Reading the Male Body," in *The Male Body*, ed. L. Goldstein (University of Michigan Press, 1994), 697.

2. Iris Young, "Throwing Like a Girl," in *Throwing Like a Girl and Other Essays in Feminist Philosophy and Social Theory* (Bloomington: Indiana University Press, 1990). The retrospective, "Throwing Like a Girl: Twenty Years Later," is in *Body and Flesh*, ed. Donn Welton (Oxford: Blackwell, 1998).

3. I count Foucault, despite disclaimers, as belonging to the phenomenological tradition. Not only was he a student of Merleau-Ponty, but reading Foucault shows many apparently deliberate inversions of Pontean claims and examples, indicating the heritage.

4. Maurice Merleau-Ponty, *Phenomenology of Perception*, trans. Colin Smith (London: Routledge and Kegan Paul, 1962).

5. See chapter 1 of this book.

6. Bordo, "Reading the Male Body," 698.

7. Michel Foucault, *The Birth of the Clinic*, trans. A. M. Sheridan Smith (New York: Vintage Books, 1973); *The Order of Things* (New York: Vintage Books, 1970); *Discipline and Punish*, trans. A. M. Sheridan Smith (New York: Pantheon Books, 1977).

8. I refer to numerous references to *the* male gaze, as if it were one.

9. In Young, *Throwing Like a Girl*.

10. Susan Bordo, "My Father the Feminist," in *Men Doing Feminism*, ed. Tom Digby (New York: Routledge, 1998).

3. Visualism in Science

1. The theses elaborated here may be found in more complete form in my *Expanding Hermeneutics: Visualism in Science* (Evanston, Ill.: Northwestern University Press, 1998).

2. The basic text that outlines a phenomenological theory of bodily active perception is Maurice Merleau-Ponty, *The Phenomenology of Perception* (New York: Humanities Press, 1962). Hubert Dreyfus's famous critique of artificial intelligence is based upon this same bodily action theory of perception and fits into a more contemporary context; see his *What Computers Can't Do* (New York: Harper Colophon Books, 1979).

3. "News Focus," *Science* 282 (4 December 1998): 1608.

4. "Sound," *Britannica Macropedia* (Encyclopaedia Britannica, 1994), vol. 27, 568–69.

5. Ibid., 569.

6. "News Focus," *Science* 281 (11 September 1998): 1597.

7. Steven Feld, "Waterfalls of Song: An Acoustemology of Place Resounding in Bosavi, Papua New Guinea," in *Senses of Place*, ed. Steven Feld and Keith H. Basso (Santa Fe, N.M.: School of American Research Press, 1996), 91–136.

8. Bettyann Kevles, *Naked to the Bone: Medical Imaging in the Twentieth Century* (New Brunswick, N.J.: Rutgers University Press, 1996), 13.

9. "Camera Obscura," *Encyclopaedia Britannica*, vol. 4 (1929), 659.

10. Ibid.

11. Daniel Boorstin, *The Discoverers* (New York: Random House, Vintage Books, 1985), 318.

12. Kevles, *Naked to the Bone*, 15.

13. Jon Darius, *Beyond Vision* (New York: Oxford University Press, 1984), 11.

14. Lee W. Bailey, "Skull's Darkroom: The Camera Obscura and Subjectivity," in *Philosophy of Technology*, ed. Paul T. Durbin (Boston: Kluwer Academic Publishers, 1989), 63–79.

15. Bruno Latour, *Science in Action* (Cambridge: Harvard University Press, 1987), 68.

4. Perceptual Reasoning

1. Andrew Pickering, *Constructing Quarks: A Sociological History of Particle Physics* (Edinburgh: Edinburgh University Press, 1984).

2. Joseph Rouse, *Knowledge as Power: Toward a Political Philosophy of Science* (Ithaca, N.Y.: Cornell University Press, 1987).

3. Sandra Harding, *The Science Question in Feminism* (Ithaca, N.Y.: Cornell University Press, 1984).

4. Don Ihde, *Instrumental Realism* (Bloomington: Indiana University Press, 1991).

5. Appendix in Edmund Husserl, *The Crisis of European Sciences and Transcendental Phenomenology* (Evanston, Ill.: Northwestern University Press, 1970).

5. You Can't Have It Both Ways

1. In Iris Young, *Throwing Like a Girl and Other Essays in Feminist Philosophy and Social Theory* (Bloomington: Indiana University Press, 1990).

2. Donna J. Haraway, *Simians, Cyborgs, and Women: Reinventing Nature* (New York: Routledge Publishers, 1991), 190.

3. "Camera Obscura." *Encyclopaedia Britannica* (1929), vol. 4, 659.

4. Ibid., 658.

5. Cited in Lee W. Bailey, "Skull's Darkroom: The *Camera Obscura* and Subjectivity," in *Philosophy of Technology*, ed. Paul T. Durbin (Boston: Kluwer Academic Publishers, 1989), 66–67.

6. Ibid., 67.

7. Ibid., 64.

8. Ibid., 68.

9. Maurice Merleau-Ponty, *The Phenomenology of Perception* (New York: Humanities Press, 1962), xi.

10. Ibid., x.

11. Ibid., 82.

12. Andrew Pickering, *Science as Practice and Culture* (Chicago: University of Chicago Press, 1992).

13. Collins quoted in ibid., 303.

14. Andrew Pickering, *The Mangle of Practice: Time, Agency, and Science* (Chicago: University of Chicago Press, 1995), 15.

15. Donna J. Haraway, *Modest-Witness@Second-Millenium. FemaleMan-Meets-OncoMouse* (New York: Routledge, 1997), 267.

16. "This Week in Science," *Science* 286 (1 October 1999): 7.

17. Don Ihde, *Technology and the Lifeworld* (Bloomington: Indiana University Press, 1990), 176.

18. "Technologies and the Quality of Life" (paper presented at the Society for Philosophy and Technology, Dusseldorf, Germany, September 1996).

6. Failure of the Nonhumans

1. Bruno Latour, *Pandora's Hope: Essays on the Reality of Science Studies* (Cambridge: Harvard University Press, 1999), 3.

2. Andrew Pickering, *The Mangle of Practice: Time, Agency, and Science* (Chicago: University of Chicago Press, 1995), xi.

3. Ibid.

4. Donna J. Haraway, *Simians, Cyborgs, and Women: The Reinvention of Nature* (New York: Routledge, 1991), 149, 152.

5. Pickering, *The Mangle of Practice*, 17.

6. Latour, *Pandora's Hope*, 16.

7. If citations and extended commentary and adaptation may be taken as indicative, then Steven Shapin and Simon Schaffer, *Leviathan and the Air-Pump: Hobbes, Boyle, and the Experimental Life* (Princeton, N.J.: Princeton University Press, 1985) became the watershed book for much contemporary science studies. Latour adopts the notion of a "modern constitution" from it in his *We Have Never Been Modern* (Cambridge: Harvard University Press, 1993), Haraway takes the notion of a "modest witness" into her scheme in *Modest-Witness@Second-Millennium* (New York: Routledge, 1997), and it was extensively cited in Pickering's *The Mangle of Practice*.

8. Shapin and Schaffer, *Leviathan and the Air-Pump*, 25.

9. Ibid., 36–37.

10. Latour, *We Have Never Been Modern*, 197.

11. I find the notion of "socialization" open to considerable doubt. To make humans and nonhumans symmetrical, Michel Callon, first, followed by Latour, attempted to take the nonhumans into a socialized process, or at least into such a vocabulary. Callon's famous scallops were "recruited," "negotiated with," gave "assent," etc. Such hyloanthropomorphism seems untoward to most philosophers.

12. Latour, *We Have Never Been Modern*, 198, 199.

13. Haraway, *Modest-Witness@Second-Millennium*, 151.

14. Pickering, *The Mangle of Practice*, 15.

15. *Technics and Praxis: A Philosophy of Technology* (Dordrecht: Reidel, 1979) was often noted as the first philosophy of technology book in North America. This work was followed, in 1990, by *Technology and the Lifeworld: From Garden to Earth* (Bloomington: Indiana University Press). Both affirmed the notion of human-technology relations as the primary units of analysis and undertook a somewhat materialist perspective upon such relations.

16. Ihde, *Technology and the Lifeworld*, 27.

17. Latour, *We Have Never Been Modern*, 179. This same illustration was used by Latour first in 1994 in "On Technical Mediation," *Common Knowledge* 3, no. 2: 29–64.

18. Latour, *We Have Never Been Modern*, 179.

19. Ibid.

20. Umberto Eco, *Foucault's Pendulum* (San Diego: Harcourt Brace Jovanovich, 1988), 3.

21. Catherine Wilson, *The Invisible World: Early Modern Philosophy and the Invention of the Microscope* (Princeton: Princeton University Press, 1992), 216.

22. Friedrich Kittler, "The Mechanized Philosopher," in *Looking after Nietzsche*, ed. Laurence Rickels (Albany: SUNY Press, 1990), 201.

23. Ibid., 195.

24. Ibid., 196.

25. Martin Heidegger, *Parmenides* (Frankfurt: Klostermann, 1982), 118–19.

26. Kittler, "The Mechanized Philosopher," 196.

27. Ibid., 197.

28. Kittler, in the article I have been citing, expands what he takes as Nietzsche's metaphysics of writing into its transformation with women taking up the roles as assistants and secretaries. Writing thus becomes "heterosexual" in this interpretation in the late nineteenth century.

29. My emphasis; cited in Michael Heim, *Electric Language: A Philosophical Study of Word Processing* (New Haven: Yale University Press, 1987), 192.

7. Prognostic Predicaments

1. Friedrich Kittler, "The Mechanized Philosopher," in *Looking after Nietzsche*, ed. Laurence Rickels (Albany: SUNY Press, 1990), 8–9.

2. Edward Tenner, *Why Things Bite Back: Technology and the Revenge of Unintended Consequences* (New York: Alfred A. Knopf, 1996), 8–9.

3. Ibid., ix.

4. Ibid.

5. Ibid., 7.

6. Bettyann Holzmann Kevles, *Naked to the Bone: Medical Imaging in the Twentieth Century* (New Brunswick, N.J.: Rutgers University Press, 1996).

7. Ibid., 48.

8. Hubert Dreyfus, *Alchemy and Artificial Intelligence* (Santa Monica, Calif.: RAND Corp., 1965); *What Computers Can't Do* (New York: Harper and Row, 1972); *What Computers Still Can't Do* (Cambridge: MIT Press, 1992).

9. "This Week in Science," *Science* 284 (11 July 1997): 190.

10. Tenner, *Why Things Bite Back*, ix.

8. Phil-Tech Meets Eco-Phil

1. Lester R. Brown, *State of the World: A World Watch Institute Report on Progress toward a Sustainable Society* (New York: Norton, 1996).

2. David Abram, *The Spell of the Sensuous* (New York: Pantheon Books, 1996).

3. Don Ihde, *Technology and the Lifeworld* (Bloomington: Indiana University Press, 1990); *Philosophy of Technology* (New York: Paragon, 1993).

Epilogue

1. Edward Tenner, *Why Things Bite Back: Technology and the Revenge of Unintended Consequences* (New York: Alfred A. Knopf, 1996), 43–44.

2. See my "Why Not Science Critics," in *Expanding Hermeneutics: Visualism in Science* (Evanston, Ill.: Northwestern University Press, 1998), 127ff.

3. Edward R. Tufte, *Visual Explanations* (Chesire, Conn.: Graphics Press, 1997), 24.

4. Nigel Henbest and Michael Marten, *The New Astronomy*, 2d ed. (New York: Cambridge University Press, 1996), 26.

5. Lincoln F. Pratson and William F. Haxby, "Panoramas of the Seafloor," *Scientific American* 276, no. 6 (June 1997): 82ff.

Index

Don Ihde is distinguished professor of philosophy at State University of New York, Stony Brook, and the author of twelve books.